SpringerBriefs in Molecular Science

Electrical and Magnetic Properties of Atoms, Molecules, and Clusters

Series Editor

George Maroulis

For further volumes:
http://www.springer.com/series/11837

Roberto Cammi

Molecular Response Functions for the Polarizable Continuum Model

Physical Basis and Quantum Mechanical Formalism

 Springer

Roberto Cammi
Department of Chemistry
University of Parma
Parma
Italy

ISSN 2191-5407   ISSN 2191-5415 (electronic)
ISBN 978-3-319-00986-5   ISBN 978-3-319-00987-2 (eBook)
DOI 10.1007/978-3-319-00987-2
Springer Cham Heidelberg New York Dordrecht London

Library of Congress Control Number: 2013942662

Printed on acid-free paper

Springer is part of Springer Science+Business Media (www.springer.com)

To Giovanna, Davide, and Paolo

Preface

Molecular response functions (MRF) are basic properties which characterize a molecular system in terms of the changes in its Quantum Mechanical (QM) observables when it is perturbed by some agent [1–5].[1] The MRF may be defined with respect to very general perturbations, which may be both physical fields and other material systems, and their use pervades many aspects of the Quantum-Chemical interpretation of chemical processes [6–15], including those occurring in solution.

A large part of the primary molecular events in chemical processes takes place in liquid solutions, or in more complex environments, and from a QM perspective the effects of the environment may be usefully expressed in terms of the effect of the solvent on the QM molecular response functions [16–24] of molecular systems involved.

In general, the description of solvation effects is an important issue in quantum chemistry. There are two main approaches, making use the first of a discrete representation of the solvent, and the second of a continuous responsive distribution.

The discrete models (QM/MM) combine a QM calculation for the molecular solute with a computer simulation (MC/MD) for the solvent, in which all the numerous degrees of freedom of the solvent molecules have to be explicitly considered.

The QM continuum solvation methods (QM/CSM) have a more simple physical and computational structure, as no explicit molecular degrees of freedom of the solvent enter into the calculation. The procedure is based on the definition of an effective Hamiltonian for the molecular solute, which is composed by the Hamiltonian of the isolated solute accompanied by a solute–solvent integral operators, with a nonlinear kernel, and describing the solute in the presence of the solvent reaction potential. The solution of the corresponding nonlinear Schrödinger equation, obtained at ab-initio QM level and with an iterative procedure, determines the properties of the molecular solute in the presence of the solvent, with a complete description of the solvent effects.

---

[1] These changes refer to a suitable reference unperturbed state of the molecular system.

There are three current approaches to continuum solvation models [25–27], according to three different approaches to the solution of the basic electrostatic problem (Poisson problem): The Generalized Born approximation, the methods based on multipolar expansions of the electrostatic potential for the analytical solution of the electrostatic problem, and the methods based on a direct numerical integration of the electrostatic problem.[2]

In the QM/GB method, the molecular charge distribution of the solute is reduced to a sum of atomic point charges, each placed within a sphere of appropriate radius, and the solvent reaction potential produced by these charges is evaluated with an approximated formula generalizing the Born expressions for a single charge within a sphere, and introduced into the Hamiltonian of the solute (for a review see [26]).

Multipolar expansions methods (MPE) have problems with the description of the solvent reaction field for solutes of irregular shape. The QM/MPE method developed by Rivail and coworkers have been for a long time limited to ellipsoidal cavities [28–29]. Mikkelsen adopts the spherical cavity [30–34], and uses the SCRF approach in combination with high-quality QM procedures for the evaluation of the molecular response functions.

Within the approaches based on the numerical integration of the electrostatic problem, the so called apparent surface charges (ASC) method, is by far computationally faster.

The first ASC method, known as Polarizable Continuum Model (PCM), has been proposed by Tomasi and coworker in 1981 [35] and since then has been continuously developed (see Tomasi et al. [27] for a recent review). Actually, there is a family of PCM methods: C-PCM [36], a version widely used, and the integral equation formalism IEF-PCM [37–39], the SCI-PCM [40], developed by K.B. Wiberg, and the recent CSC-PCM [41] developed by G. Scalmani and M. J. Frisch. Other ASC methods amply used are COSMO [42], developed by A. Klamt and the SVEP [43] developed by D. Chipman.

The PCM model contains the largest variety of extensions for the calculation of the properties for the ground and excites states of molecular systems in solution [44–49] and these extensions have been accomplished at HF level and at various QM electron correlation methods [27, 50–57]. There are also version based on semi-empirical QM methods: we quote here only those based on ZINDO [58].

The aim of this book is to present the basic aspects of the molecular response function theory for molecular systems in solution described with the Polarizable Continuum Model, giving special emphasis both to the physical basis of the theory and to its quantum chemical formalism. The QM formalism will be presented in the form of the coupled-cluster theory, as it is the most recent and less known formulation for the QM calculation of molecular properties within the PCM

---

[2] Note that we are here speaking about the electrostatic problem only, discarding the non-electrostatic components of the solute–solvent interaction. For a detailed discussion of these interactions See Ref. [27].

model. Although the book is focused on the description based on the PCM many of the topics that will be treated can be easily translated into the discrete solvation models, and also in other forms of the Continuum Models.

The book is organized as follows. Chapter 1 introduces the basic definitions of the PCM model. The focus is on the electrostatic problem for the determination of the solute–solvent integral operator and on the associated nonlinear QM problem, which is exemplified at the Hartree-Fock and at the Coupled-cluster methods. Chapter 2 considers the derivatives of the basic energy functional of the PCM model with respect to perturbing fields, the related generalized Helmann-Feynman theorem, and their analytical evaluation at the coupled-cluster level. Chapter 3 presents the general aspects response theory for the PCM model. They include the descriptions of the non-equilibrium solvation, the variational formulation of the time-dependent nonlinear QM problem, and the connection of the molecular response functions with the corresponding macroscopic counterparts. Chapter 4 presents the properties of the excited states of molecular solutes as determined from the transition properties of the coupled-cluster linear response function described in Chap. 3. Finally, the Appendix collects several interesting details concerning the physics and the QM formalism of the PCM model, including also a formulation of the corresponding molecular electronic virial theorem.

Parma, March 2013   Roberto Cammi

References

1. D.N. Zubarev, *Nonequlibrium Statistical Thermodynamics* (Consultant Bureau, Plenum, New York, 1974)
2. J. Lindenberg, Y. Öhrn, *Propagators in Quantum Chemistry* (Academic Press, London, 1973)
3. P. Jørgensen, J. Simons, *Second Quantization Based Methods in Quantum Chemistry* (Academic Press, New York, 1981)
4. J. Olsen, P. Jørgensen, in *Modern Electronic Structure Theory*, vol. 2, ed. by D. Yarkony (World Scientific, Singapore, 1995), p. 857
5. T. Helgaker, S. Coriani, P. Jørgensen, K. Kristensen, J. Olsen, K. Ruud, Chem. Rev. **112**, 543 (2012)
6. A.D. Buckingham, Adv. Chem. Phys. **12**, 107 (1964)
7. D.P. Craing, T. Thirunamachandran, *Molecular Quantum Electrodynamics* (Academic Press, New York, 1984)
8. L. Barron, *Molecular Light Scattering and Optical Activity* (Cambridge University Press, Cambridge, UK, 1983)
9. D. Bishop, Rev. Mod. Phys. **62**, 343 (1990)
10. S. Mukamel, *Principles of Non-Linear Optical Spectroscopy* (Oxford University Press, New York, 1995)
11. G. Parr, R.G. Pearson, J. Am. Chem. Soc. **105**, 7512 (1983)
12. G. Parr, W.T. Yang, J. Am. Chem. Soc. **106**, 4049 (1984)
13. J. Simons, K.D. Jordan, Chem. Rev. **87**, 535 (1987)
14. J.V. Ortiz, Adv. Quantum Chem. **35**, 233 (1999)

15. J. Simons, Adv. Quantum Chem. **50**, 213 (2005)
16. D.M. Bishop, Int. Rev. Phys. Chem. **13**, 21 (1994)
17. K.V. Mikkelsen, Y. Luo, H. gren, P. Jørgensen, J. Chem. Phys. **100**, 8240 (1994)
18. J. Tomasi, R. Cammi, B. Mennucci, Int. J. Quantum Chem. **75**, 767 (1999)
19. Y. Luo, P. Norman, P. Macak, H. Ågren, Phys. Rev. B **61**, 3060 (2000)
20. B. Champagne, D.M. Bishop, Adv. Chem. Phys. **126**, 41 (2003)
21. L. Jensen, K.O. Sylvester-Hvid, K.V. Mikkelsen, P.O. Strand, J. Phys. Chem. A **107**, 2270 (2003)
22. J. Holtmann, E.Walczuk, M. Dede, C. Wittenburg, J. Heck, G. Archetti, R. Wortmann, H.G. Kuball, Y.H. Wang, K. Liu, Y. Luo, J. Phys. Chem. B **112**, 14751 (2008)
23. L. Ferrighi, L. Frediani, C. Cappelli, P. Salek, H. Ågren, T. Helgaker, K. Ruud, Chem. Phys. Lett. **425**, 267 (2006)
24. C.B. Nielsen, O. Christiansen, K.V. Mikkelsen, J. Kongsted, J. Chem. Phys. **126**, 154112 (2007)
25. J. Tomasi, M. Persico, Chem. Rev. **94**, 2027 (1994)
26. C.J. Cramer, D.G. Truhlar, Chem. Rev. **99**, 2161 (1994)
27. J. Tomasi, B. Mennucci, R. Cammi, Chem. Rev. **105**, 2999 (2005)
28. J.L. Rivail, D. Rinaldi, Chem. Phys. **18**, 25 (1976)
29. D. Rinaldi, J.L. Rivail, Theor. Chim. Acta. **32**, 57 (1978)
30. K.V. Mikkelsen, H. Agreen, H.J.A. Jensen, T. Helgaker, J. Chem. Phys. **89**, 3086 (1988)
31. K.V. Mikkelsen, P. Jorgensen, H.J.A. Jensen, J. Chem. Phys. **100**, 6597 (1994)
32. K.V. Mikkelsen, A. Cesar, H. Agreen, H.J.A. Jensen, J. Chem. Phys. **103**, 9010 (1995)
33. O. Christiansen, K.V. Mikkelsen, J. Chem. Phys. **110**, 1365 (1999)
34. O. Christiansen, K.V. Mikkelsen, J. Chem. Phys. **110**, 8348 (1999)
35. S. Miertuš, E. Scrocco, J. Tomasi, Chem. Phys. **55**, 117 (1981)
36. M. Cossi, V. Barone, J. Phys. Chem. A **102**, 1995 (1998)
37. E. Cancès, B. Mennucci, J. Tomasi, J. Chem. Phys. **107**, 3092 (1997)
38. B. Mennucci, E. Cancès, J. Tomasi, J. Phys. Chem. B **101**, 10506 (1997)
39. E. Cancès, B. Mennucci, J. Math. Chem. **23**, 309 (1998)
40. J.B. Foresman, T.A. Keith, K.B. Wiberg, J. Snoonian, M.J. Frisch, J. Phys. Chem. **100**,16098 (1996)
41. G. Scalmani, M.J. Frisch, J. Chem. Phys. **132**, 114110 (2010)
42. A. Klamnt, G.J. Schürmann, J. Chem. Soc. Perkin Trans. II, 799 (1999)
43. D.M. Chipman, J. Chem. Phys. **106**, 10194 (1997)
44. R. Cammi, M. Cossi, B. Mennucci, J. Tomasi, J. Chem. Phys. **105**, 10556 (1996)
45. J. Tomasi, R. Cammi, B. Mennucci, Int. J. Quantum Chem. **75**, 767 (1999)
46. J. Tomasi, R. Cammi, B. Mennucci, C. Cappelli, S. Corni, Phys. Chem. Chem. Phys. **4**, 5697 (2002)
47. R. Cammi, L. Frediani, B. Mennucci, K. Ruud, J. Chem. Phys. **119**, 5818 (2003)
48. J. Tomasi, B. Mennucci, R. Cammi, in *Handbook of Molecular Physics and Quantum Chemistry*, vol. 3, ed. by S. Wilson (Wiley, New York, 2003), pp. 299–328
49. R. Cammi, B. Mennucci, J. Tomasi, in *Computational Chemistry: Review of Current Trends*, vol. 8, ed. by J. Leszczynski (World Scientific, Singapore, 2003), pp. 1–79
50. R. Cammi, J. Chem. Phys. **131**, 164104 (2009)
51. M. Caricato, G. Scalmani, G.W. Trucks, M.J. Frisch, J. Phys. Chem. Lett. **1**, 2369 (2010)
52. M. Caricato, G. Scalmani, M.J. Frisch, J. Chem. Phys. **134**, 244113 (2011)
53. M. Caricato, J. Chem. Phys. **135**, 074113 (2011)
54. R. Cammi, Int. J. Quantum Chem. **110**, 3040 (2010)
55. R. Cammi, R. Fukuda, M. Ehara, H. Nakatsuji, J. Chem. Phys. **133**, 024104 (2010)
56. R. Fukuda, M. Ehara, H. Nakatsuji, R. Cammi, J. Chem. Phys. **134**, 104109 (2011)
57. R. Cammi, Int. J. Quantum Chem. **112**, 2547 (2012)
58. M. Caricato, B. Mennucci, J. Tomasi, J. Phys. Chem. A. **104**, 5631 (2004)

Acknowledgments

The author would like to thank his mentor Prof. J. Tomasi, and the colleagues: C. Cappelli, S. Corni, M. Cossi, L. Frediani, M. Caricato, B. Mennucci, and C. Pomelli, who have contributed to the developments of the Molecular Response Functions method for the Polarizable Continuum Model. The author also thanks Prof. G. Maroulis for the invitation to contribute with a volume to this Springer Series, and Gaussian Inc. for support.

Contents

Acronyms

/

AO	Atomic Orbitals basis functions
ASC	Apparent Surface Charges
BD	Coupled cluster Bruckner double method
CC	Coupled cluster method
CCSD	CC Single-double method
CISD	Configuration Interaction single-double method
COSMO	Conductor Screening Model
C-PCM	Conductor like Polarizable Continuum Model
CSM	Continuum Solvation Model
DFT	Density Functional Theory
EFISHG	Electric Field Induced Second Harmonic Generation
EOM-CC	Equation of Motion Coupled cluster method
HF	Hartree-Fock method
H-F	Helmann-Feynman theorem
IEF-PCM	Integral Equation Formalism version of the PCM model
LR	Linear Response method
MC	Monte Carlo simulation
MCSCF	Multiconfigurational self-consistent field method
MD	Molecular dynamics
MM	Molecular Mechanics
MO	Molecular orbitals
MP2	2nd order Møller-Plesset method
PCM	Polarizable Continuum Model
PCM-CC	PCM Coupled Cluster method
PCM-CC-LR	Polarizable Continuum Model Coupled Cluster Linear Response method
PCM-CC-PTE	PCM-CC with Perturbation Theory on Energy
PCM-CCSD	PCM-CC single-double method
PCM-CCSD-PTE	PCM-CC single-double method with Perturbation theory on energy
PCM-EOM-CC	Polarizable Continuum Model EOM-CC method
PCM-HF	PCM Hartree-Fock method
PCM-LR	Polarizable Continuum Model Linear Response method
PCM-SACCI	Polarizable Continuum Model SACCI

PTE	Perturbation Theory on Energy
QM	Quantum mechanics
SAC	Symmetry Adapted Cluster
SAC-CI	Symmetry adapted cluster CI
SCF	Self-Consistent Field method
SCRF	Self Consistent Reaction Field method
SHG	Second Harmonic Generation
SVEP	Surface Volume Explicit Polarization model
TDNLSE	Time-dependent non-Linear Schrödinger equation
EVT	Electronic Virial Theorem

Chapter 1
The PCM Model

Abstract This chapter introduces to the basic definitions of the PCM model for a molecular solute. The basic electrostatic problem for the determination of the solute-solvent interaction is described within the Integral Equation Formalism (IEF-PCM), and the QM problem associated to the effective Hamiltonian of the molecular solute is formulated in terms of a basic energy functional which has the thermodynamic status of a free-energy for the entire solute-solvent system. The QM problems for the molecular solute is exemplified at the Hartree-Fock and at the coupled-cluster level methods.

1.1 The PCM: Basic Concepts and Definitions

The PCM [1] is an ab-initio quantum mechanical (QM) method to describe the solvent effect on the properties of a molecular systems. In its basic version, the PCM model represents the solvent as a homogeneous and infinite dielectric medium having the same dielectric permittivity ε of the pure solvent, and hosting the molecular solute within an accurately modeled void cavity.

The effective electronic Hamiltonian[1] for the molecular solute M is [2]:

$$H = H^o + V(\Psi) \tag{1.1}$$

where H^o is the electronic Hamiltonian of the isolated molecule and $V(\Psi)$ is an effective solute-solvent interaction operator. In the basic PCM, $V(\Psi)$ is expressed in terms the solvent reaction potential (see V_σ in next Eq. (1.2)) generated by the dielectric medium polarized by the whole (electronic and nuclear) charge distribution $\rho_M = \rho_M^e + \rho_M^n$ of the molecular solute. The argument of the solute-solvent

---

[1] We assume the usual Born-Oppenheimer non relativistic approximation and H is the electronic Hamiltonian for the molecule solute with fixed nuclei coordinates **R**.

R. Cammi, *Molecular Response Functions for the Polarizable Continuum Model*,
Springer Briefs in Electrical and Magnetic Properties of Atoms, Molecules, and Clusters,
DOI: 10.1007/978-3-319-00987-2_1, © The Author(s) 2013

interaction potential $V(\Psi)$ indicates its dependence on the solute wavefunction Ψ via the electronic charge density $\rho_M^e(\mathbf{r})$.

As other QM continuum models, the PCM model requires the solution of two coupled problems: an electrostatic classical problem for the determination of the solvent reaction potential V_σ induced by the total charge distribution ρ_M; and a quantum mechanical problem for the determination of the wavefunction Ψ of the solute described by the effective QM Hamiltonian (1.1). The two problems are nested and they must be solved simultaneously.

With respect to other QM continuum models, the PCM method represents of the interaction operator $V(\Psi)$ (i.e. of the solvent reaction potential V_σ) in terms of an apparent surface charge (ASC) charge distribution σ spread on the boundary Γ of the cavity (**C**) hosting the solute M.

1.1.1 The Definition of Cavity

The cavity is a constitutive component in all the continuum methods [2, 3]. In the PCM model the cavity may be accurately modeled on the shape of the molecular solute M. The basic PCM cavity is defined as a set of interlocking spheres centered on the nuclei of atoms of M, with radii related to the corresponding atomic van der Waals radii.[2] Actually, the PCM model uses several variants of the basic cavity, which may introduce additional spheres [4, 5], to take into account of the portions of space not occupied by the charge distribution of molecular solute but non accessible or excluded to the solvent molecules.

1.1.2 The Electrostatic Problem

The physics of the solute-solvent interaction potential $V(\Psi)$ of Eq. (1.1) is simple. The total charge distribution ρ_M of the molecular solute polarizes the dielectric medium, which in turn becomes source of an additional electrostatic potential V_σ (the *solvent reaction potential*) for the electrons and nuclei of the molecular solute. $V(\Psi)$ can be written as

$$V(\Psi) = \sum_{i=1}^{N} -V_\sigma(\mathbf{r}) + \sum_{\alpha=1}^{M} Z_\alpha V_\sigma(\mathbf{R}_\alpha) \qquad (1.2)$$

for a system of N electrons with coordinates $\mathbf{r}_i$ and M nuclei with fixed coordinates $\mathbf{R}_\alpha$.

---

[2] The radius of the atomic spheres is an adjustable parameter of the PCM model.

The solvent reaction potential V_σ in Eq. (1.2) is determined by solving the classical electrostatic Poisson equation which governs total electrostatic potential $V = V_M + V_\sigma$ (V_M is the electrostatic potential produced by the electronic and nuclear charge distribution of the solute). The Poisson problem has the form of a partial differential equations with domain in the whole three-dimensional space [2][3]

$$\begin{aligned} -\nabla^2 V(\mathbf{r}) &= 4\pi\rho_M(\mathbf{r}) \quad \mathbf{r} \subseteq \mathbf{C} \\ -\nabla^2 V(\mathbf{r}) &= 0 \quad \mathbf{r} \nsubseteq \mathbf{C} \end{aligned} \tag{1.3}$$

with the additional boundary conditions on the total electrostatic potential V at the infinity, and across of the cavity surface Γ.[4]

The solve the Poisson problem (1.3), the PCM model exploits an integral representation[5] of the solvent reaction potential V_σ:

$$V_\sigma(\mathbf{r}) = \int_\Gamma \frac{\sigma(\mathbf{s})}{|\mathbf{r} - \mathbf{s}|} d\mathbf{s} \quad \mathbf{s} \subset \Gamma \tag{1.4}$$

where $\sigma(\mathbf{s})$ is an Apparent Surface Charges (ASC) distribution spread on the cavity surface Γ.[6]

In computational practice, the ASC distribution $\sigma(\mathbf{s})$ in (1.4) is discretized in terms of a finite set of point charges $\{q(\mathbf{s}_k)\}$ uniformly distributed on the cavity surface Γ (at the positions $\mathbf{s}_k$), and the Poisson problem is transformed into a matrix equation (PCM equation) for the unknown apparent charges $\{q(\mathbf{s}_k)\}$ [2].

There are several variants of the PCM matrix equation (see Ref. [2]). The most general form is based on the Integral Equation Formalism (PCM-IEF) [6] and, for

---

[3] $\mathbf{r} \subseteq \mathbf{C}$ and $\mathbf{r} \nsubseteq \mathbf{C}$ denote, respectively, points inside and outside of the cavity $\mathbf{C}$

[4] The boundary conditions for the total electrostatic potential V are: at the infinity

$$\begin{aligned} \lim_{r \to \infty} r V(\mathbf{r}) &= 0 \\ \lim_{r \to \infty} r^2 \nabla V(\mathbf{r}) &= 0 \end{aligned}$$

and across the boundary Γ of the cavity

$$\begin{aligned} V_{in}(\mathbf{s}) &= V_{out}(\mathbf{s}) \\ \frac{\partial V_{in}(\mathbf{s})}{\partial \mathbf{n}} &= \epsilon \frac{\partial V_{out}(\mathbf{s})}{\partial \mathbf{n}} \end{aligned}$$

with $\mathbf{s} \subset \Gamma$

[5] The PCM approach is related to the Boundary Elements Methods (BEM), a numerical method to solve the solve complex partial differential equations with domain in the whole three-dimensional space via numerical integration of integral equation with domain in two-dimensional boundary surfaces. For more information on the BEM see the Website http://www.boundary-elements-methods.com.

[6] The Poisson problem is then converted into a single integral equation with domain on the cavity boundary Γ, for the unknown ASC distribution $\sigma(\mathbf{s})$:

$$\left(2\pi \frac{\epsilon + 1}{\epsilon - 1} + D^* \right) \sigma(\mathbf{s}) = \frac{\partial V_M(\mathbf{s})}{\partial \mathbf{n}_s} \quad \mathbf{s} \subset \Gamma \tag{1.5}$$

the case of a homogeneous dielectric, it determines the polarization charges from the electrostatic potential produced by the solute at the cavity cavity surface.

More specifically, the PCM matrix equation may be written as

$$\bar{Q}(\Psi) = T < \Psi|V|\Psi > \tag{1.6}$$

where:

- $\bar{Q}$ is a vector column collecting the polarization charges $\{q(s_k)\}$:

$$[\bar{Q}]_K = q(s_k)$$

- T is a matrix which represents the responsive polarization of the solvent, depending on its static dielectric permittivity ϵ_0 of the medium and on the geometry Γ of the cavity hosting the solute,
- V is vector collecting the electrostatic potential operator (2.9) of the solute at positions s_k:

$$[V]_K = V_M(s_K)$$

with the electrostatic potential operator $\hat{V}_M(s)$, sum of its electronic and nuclear contributions[7]:

$$\hat{V}_M(s) = \hat{V}_M(s)_{el} + \hat{V}_M(s)_{nuc} \tag{1.7}$$

The solvent polarization charges $\bar{Q}(\Psi)$ of Eq. (1.6) may rewritten as an expectation value [7]

$$\bar{Q}(\Psi) = < \Psi|Q|\Psi > \tag{1.8}$$

where Q is the apparent charge operator defined as

$$Q = T \cdot V \tag{1.9}$$

In terms of the polarization charges $\bar{Q}$ (1.8) the solute-solvent interaction operator $V(\Psi)$ may be written as:

(Footnote 6 continued)

Here:

- D^* is an integral operator [6] defined as

$$D^*\sigma(s) = \int_\Gamma \left(n_s \cdot \nabla \frac{1}{|s - s'|}\right) \sigma(s')ds')$$

- $\partial V_M(s)/\partial n_s$ is directional derivatives, normal to the cavity surface, of the total molecular electrostatic potential.

[7] The electronic and nuclear electrostatic potential operators are: $\hat{V}_M((s))_{el} = \sum_i^N \frac{-1}{|r_i - s|}$ and $\hat{V}_M((s))_n = \sum_\alpha^N \frac{-1}{|R - s|}$.

$$V(\Psi) = < \Psi|\mathbf{Q}|\Psi > \cdot\mathbf{V}$$

and the effective Hamiltonian (1.1) becomes

$$H = H^o + < \Psi|\mathbf{Q}|\Psi > \cdot\mathbf{V} \tag{1.10}$$

1.1.3 The Quantum Mechanical Problem

An effective approach to the QM problem for the PCM model is based on the definition of a suitable basic QM energy functional [2]:

$$G = < \Psi|H^o + \frac{1}{2} < \Psi|\mathbf{Q}|\Psi > \cdot\mathbf{V}|\Psi > \tag{1.11}$$

, which does not correspond to the expectation valued of the effective Hamiltonian (1.1) $E = < \Psi|H^o + < \Psi|\mathbf{Q}|\Psi > \cdot\mathbf{V}|\Psi >$ as a consequence of the dependence on the solute wavefunction Ψ of the solute-solvent operator[8] [8].

The energy functional G of Eq. (1.11) is also called free-energy functional because it has the thermodynamical status of the free-energy of the whole solute-solvent systems. More specifically, it refers to a reference state given by the non interacting electron and nuclei of the molecular solute, at rest, and by the unperturbed, pure solvent at the standard thermodynamic conditions of temperature and pressure.

For the QM energy functional (1.11) hold the following properties:

- The stationarity condition

$$\delta G = \delta < \Psi|H^o + \frac{1}{2} < \Psi|\mathbf{Q}|\Psi > \cdot\mathbf{V}|\Psi >= 0, \tag{1.12}$$

· supplemented by the usual constraint of normalization on the wavefunction $< \Psi|\Psi >= 1$ leads to the time-independent non-linear Schrödinger equation for the effective Hamiltonian (1.10)

$$[H^o + < \Psi|\mathbf{Q}|\Psi > \cdot\mathbf{V}]|\Psi >= E|\Psi > \tag{1.13}$$

All the information on the electronic structure and properties in the stationary states of a molecular solute can be determined by the solution of Eq. (1.13).
- The free-energy functional $G(\mathbf{R})$ is defined at a given fixed nuclei position of the molecular solute. Therefore, using the standard Born-Oppenheimer approach, the free energy functional $G(\mathbf{R})$ may be considered as a function of the nuclear

---

[8] We note that the difference between the free-energy functional (1.11) and the expectation value of the effective Hamiltonian (1.1) corresponds to virtual work needed to polarize the solvent in the absence of the molecular solute.

coordinates, and the surface formed by $G(\mathbf{R})$ as function of the nuclear coordinate is the potential energy surface (PES) for the nuclei of the molecular solute.

- The free-functional $G(\mathbf{R})$ is also the basic energy quantity for the determination of the time-independent properties of the molecular solutes, as will be discussed in Chap. 3.

In the following subsections we will consider the electronic wavefunction of the molecular solute at the Hartree-Fock level and at the coupled-cluster level of the QM theory.

1.1.4 The PCM Hartree-Fock Equations

In a N-electron system with spin-orbitals expanded over a set of atomic orbitals (AO) $\{\chi_\mu, \chi_\nu, \ldots\}$, the free-energy functional G (2.17) may be written as [8]:

$$G_{HF} = \sum_{\mu\nu} P_{\mu\nu}^{HF}(h_{\mu\nu} + j_{\mu\nu}) + \frac{1}{2}\sum_{\mu\nu\lambda\sigma} P_{\mu\nu}^{HF} P_{\lambda\sigma}^{HF}\left[\langle\mu\lambda||\nu\sigma\rangle + \mathscr{B}_{\mu\nu,\lambda\sigma}\right] + \tilde{V}_{NN}$$

$$(1.14)$$

where $h_{\mu\nu}$ are the matrix elements, in the AO basis, of the one-electron core operator, $\langle\mu\lambda||\nu\sigma\rangle$ are the antisymmetrized combination of regular two-electron repulsion integrals (ERIs) and $P_{\mu\nu}^{HF}$ indicates the elements of the Hartree-Fock density matrix in the AO basis. The matrix elements $j_{\mu\nu}$, and $\mathscr{B}_{\mu\nu,\lambda\sigma}$, representing the solute-solvent interactions within the PCM-Fock operator, follow from the partition of the solvent reaction potential into two components, one related to the solvent polarization induced by the nuclei charge distribution of the molecular solute, and the other on the solvent polarization due to the corresponding electronic charge distribution [9] More specifically, the one-particle AO integrals $j_{\mu\nu}$ and the pseudo two-electrons

---

[9] The partition of the electrostatic potential operator of Eq. (1.7), with the consequent partition of the vector operator $\mathbf{V} = \mathbf{V}_e + \mathbf{V}_n$, leads a parallel partition of the apparent charges $\bar{\mathbf{Q}}(\Psi)$ into electronic and nuclear contributions:

$$\bar{\mathbf{Q}}(\Psi) = \bar{\mathbf{Q}}(\Psi)_e + \bar{\mathbf{Q}}_n \qquad (1.15)$$

with

$$\bar{\mathbf{Q}}(\Psi)_e = <\Psi|\mathbf{Q}_e|\Psi>, \qquad \mathbf{Q}_e = \mathbf{T}\mathbf{V}_e$$
$$\bar{\mathbf{Q}}_n = \mathbf{Q}_n = \mathbf{T}\mathbf{V}_n$$

where $\mathbf{V}_e$ and $\mathbf{V}_n$ are, respectively the electronic and nuclear contribution to the vectorial operator $\mathbf{V}$.

integrals $\mathscr{B}_{\mu\nu,\lambda\sigma}$ represent, respectively, the interactions with the nuclear and with the electronic components of the ASC charges. The solvent integrals $j_{\mu\nu}$, $\mathscr{B}_{\mu\nu,\lambda\sigma}$ may be expressed in the following form [8]

$$j_{\mu\nu} = \mathbf{v}_{\mu\nu} \cdot \mathbf{Q}_{nuc} \tag{1.16}$$

$$\mathscr{B}_{\mu\nu,\lambda\sigma} = \mathbf{v}_{\mu\nu} \cdot \mathbf{q}_{\lambda\sigma} \tag{1.17}$$

where $\mathbf{v}_{\mu\nu}$ is a vector collecting the AO integrals of the electrostatic potential operator evaluated at the positions $\mathbf{s}_k$ of the ASC charges, $< \chi_\mu | -1/|\mathbf{r} - \mathbf{s}_k| \chi_\nu >$; $\mathbf{Q}_{nuc}$ is the vector collecting the ASC charges produced by the nuclear charge distribution; $\mathbf{q}_{\lambda\sigma}$ is a vector collecting the apparent charges produced by the elementary charge distribution $\chi_\lambda^*(\mathbf{r})\chi_\sigma(\mathbf{r})$.[10]

The last term of Eq. (1.14), $\tilde{V}_{NN}$ is the nuclei–nuclei interaction contribution

$$\tilde{V}_{NN} = V_{NN} + 1/2\mathbf{V}_{nuc} \cdot \bar{\mathbf{Q}}_{nuc}$$

where V_{NN} is the nuclear repulsion energy and the second term of the right side is electrostatic interaction between the nuclei and the polarization charges induced by the nuclei itself.

Requiring that the free-energy function (1.14) be stationary ($\delta G_{HF} = 0$) with respect the variation of MO expansion coefficients, with the usual ortho-normality constraints on the MO, we obtain the PCM-HF equations:

$$\sum_\nu \left(f_{\mu\nu}^{PCM} - \epsilon_p S_{\mu\nu} \right) c_{\nu p} = 0 \tag{1.18}$$

where $S_{\mu\nu}$ and $f_{\mu\nu}^{PCM}$ are, respectively the matrix elements of the overlap matrix and of the PCM Fock matrix, in the AO basis, and ϵ_p and $c_{\nu p}$ are, respectively, the orbital energy and the expansion coefficients of the p MO; the PCM Fock matrix elements are given by:

$$f_{\mu\nu}^{PCM} = (h_{\mu\nu} + j_{\mu\nu}) + \sum_{\lambda\sigma} P_{\lambda\sigma}^{HF} \left[\langle \mu\lambda||\nu\sigma \rangle + \mathscr{B}_{\mu\nu,\lambda\sigma} \right] \tag{1.19}$$

---

[10] In terms of the PCM-IEF Eq. (1.6), the polarization charge vector $\mathbf{q}_{\lambda\sigma}$ is given by

$$\mathbf{q}_{\lambda\sigma} = \mathbf{T}\mathbf{V}_{\lambda\sigma}$$

where $\mathbf{v}_{\lambda\sigma}$ is a vectors collecting the electrostatic potential produced by the elementary charge distribution $\chi_\lambda^*(\mathbf{r})\chi_\sigma(\mathbf{r})$, at the positions of the ASC charges.

1.2 The Coupled-Cluster Theory for PCM

The PCM coupled-cluster (PCM-CC) theory [9–13] introduces an explicit description of the coupling between the electron-correlation (dynamic) of the molecular solute and the solute reaction potential. The electron-correlation modifies the charge distribution ρ_M of the solute. The changes in charge distribution ρ_M modify the solvent reaction potential V_σ, which in turn influence the electron-correlation. If the coupling between the dynamical electronic correlation of the solute and the polarization of the solvent is neglected the dynamic electron-correlation of the molecular solutes is evaluated in the presence of the fixed Hartree-Fock solvent reaction potential. This approximated form of the PCM-CC theory is denoted with the acronym PTE (i.e. Perturbation Theory on the Energy) which derives from a many-body perturbation analysis of the solute-solvent interaction [14].

The coupled-cluster wave function is defined by the exponential ansatz [15–18]

$$|CC> = e^T |HF> \qquad (1.20)$$

where $|HF>$ is the single determinant Hartree-Fock state of the molecular solute, and the cluster operator T is given as a sum of all possible excitation operators over the N electrons

$$T = T_1 + T_2 + \cdots + T_N; \qquad T_n = \frac{1}{(n!)^2} \sum_{aibj\dots} t_{ij\dots}^{ab\dots} a_a^\dagger a_i a_b^\dagger a_j \dots \qquad (1.21)$$

weighted by the amplitude t_i^a, t_{ij}^{ab}, etc. The excitation operators are here represented as products of second quantization electron creation $(a_i^\dagger, a_b^\dagger)$ and annihilation operators (a_i, a_b). As usual, indexes $(i, j, k, \dots)$ and $(a, b, c, \dots)$ denote, respectively, occupied and vacant spin orbitals MO, while $(p, q, r, \dots)$ denote general spin orbitals.

At the coupled-cluster level, the PCM energy functional may be written as [9]:

$$\Delta G_{CC} = <HF|(1+\Lambda)e^{-T}\left[H(0)_N + \frac{1}{2}\bar{\mathbf{Q}}_N(T,\Lambda)\cdot\mathbf{V}_N\right]e^T|HF> \quad (1.22)$$

where

- Λ is a de-excitation operator,

$$\Lambda = \Lambda_1 + \Lambda_2 + \cdots + \Lambda_N; \qquad \Lambda_n = \frac{1}{(n!)^2} \sum_{ijkabc\dots} \lambda_{abc\dots}^{ijk\dots} a_i^\dagger a_a a_j^\dagger a_b a_k^\dagger a_c \dots$$

$$(1.23)$$

- $H(0)_N$ is the normal ordered form of Hamiltonian of the solute in the presence of the apparent charges determined by the Hartree-Fock reference function

$(|HF >)$, defined as:

$$H(0)_N = H_N^o + \bar{\mathbf{Q}}(HF) \cdot \mathbf{V}_N \tag{1.24}$$

where H_N^o is the normal ordered Hamiltonian of the isolated molecule, and $\bar{\mathbf{Q}}(HF) =< HF|\mathbf{Q}|HF >$ collects the Hartree-Fock polarization charges, and $\mathbf{V}_N$ is a normal ordered vector collecting the electrostatic potential operators of the molecular solute evaluated at the positions of the ACS charges (see Table 1.1).

- $\bar{\mathbf{Q}}_N$ is the coupled-cluster expectation value for the apparent charge operator $\mathbf{Q}_N$ [9]

$$\bar{\mathbf{Q}}_N(T, \Lambda) =< HF|(1 + \Lambda)e^{-T}\mathbf{Q}_N e^T|HF > \tag{1.25}$$

being $\mathbf{Q}_N$ the normal ordered operator of the polarization charges (see Table 1.1).

The stationary condition of the PCM-coupled-cluster functional $\Delta G_{CC}(\Lambda, T)$ (1.22) with respect to the Λ amplitudes leads to equations for the T amplitudes

$$\frac{\partial \Delta G_{CC}}{\partial \lambda_{ab...}^{ij...}} =< HF|\tau_p^\dagger e^{-T} H_N e^T|HF >= 0 \tag{1.26}$$

where $\tau_p^\dagger$ is the adjoint of an elementary excitation operator $\tau_p = a_a^\dagger a_i a_b^\dagger a_j \ldots$, and H_N is the effective Hamiltonian of the molecular solute:

$$H_N = H(0)_N + \bar{\mathbf{Q}}_N(T, \Lambda) \cdot \mathbf{V}_N \tag{1.27}$$

being $H(0)_N$ the normal ordered form of Hamiltonian of the solute in the presence of the frozen HF polarization charges [9], and $\bar{\mathbf{Q}}_N \cdot \mathbf{V}_N$ the coupled-cluster component of the solvent reaction potential.

Equation (1.26) shows that the T amplitude equations correspond to the projection of the coupled-cluster Schrödinger equation for the molecular solute

$$[H(0)_N + \bar{\mathbf{Q}}_N(\Lambda, T) \cdot \mathbf{V}_N]e^T|HF >= \Delta E_{CC}e^T|HF > \tag{1.28}$$

Table 1.1 Definition of the normal ordered operators $H(0)_N$, $\mathbf{Q}_N$, and $\mathbf{V}_N$

$\bar{\mathbf{V}}_{HF} =< HF|\mathbf{V}|HF >$	$\bar{\mathbf{Q}}_{HF} =< HF|\mathbf{Q}|HF >$
$\mathbf{V}_N = \mathbf{V} - \bar{\mathbf{V}}_{HF}$	$\mathbf{Q}_N = \mathbf{Q} - \bar{\mathbf{Q}}_{HF}$
$H_N^o = H^o - < HF|H^o|HF >$	

H^o is the Hamiltonian operator of the isolated molecule, $\mathbf{Q}$ and $\mathbf{V}$ are, respectively, the apparent charge operator and the molecular electrostatic potential operator

where the correlation energy ΔE_{CC} is given by

$$\Delta E_{CC} =< HF|(1 + \Lambda)e^{-T}\left[H(0)_N + \bar{\mathbf{Q}}_N(\Lambda, T) \cdot \mathbf{V}_N\right]e^T|HF > \quad (1.29)$$

Conversely,[11] the stationary conditions of $\Delta G_{CC}(\Lambda, T)$ with respect to the T amplitudes leads to equations for the Λ amplitudes:

$$\frac{\partial \Delta G_{CC}}{\partial t_{ij\ldots}^{ab\ldots}} =< HF|(1 + \Lambda)e^{-T}[H_N, \tau_p]e^T|HF >= 0 \quad (1.30)$$

The Λ Eq. (1.30) are coupled with the T Eq. (1.26) as they involves effective Hamiltonian H_N (1.27) which depends on both T and Λ amplitudes, and vice-versa. Therefore Eqs. (1.30) and (1.26) must be solved simultaneously.

The PCM coupled-cluster theory has been presented at the coupled-cluster single and double (CCSD) excitation level approximation [9, 11], at the Brueckner doubles (BD) coupled-cluster level [12], and within the symmetry adapted cluster (SAC) method [10].

1.2.1 The PCM-CC-PTE Approximation

In the PTE approximation, which neglect effect of the electron-correlation on the solvent reaction potential, the PCM free-energy functional becomes

$$\Delta G_{CC}^{PTE} =< HF|(1 + \Lambda)e^{-T}H(0)_N e^T|HF > \quad (1.31)$$

and the stationary conditions for the T and Λ amplitudes become:

$$< HF|\tau_p e^{-T}H(0)_N e^T|HF >= 0 \quad (1.32a)$$

$$< HF|(1 + \Lambda)e^{-T}H(0)_N, \tau_p]e^T|HF >= 0 \quad (1.32b)$$

where the T amplitude Eq. (1.32a) correspond to the projection of the coupled-cluster Schrödinger equation for the molecular solute

$$H(0)_N e^T|HF >= \Delta E_{CC}^{PTE} e^T|HF > \quad (1.33)$$

where $\Delta E_{CC}^{PTE} = \Delta G_{CC}^{PTE}$ is the correlation energy.

---

[11] The correlation energy ΔE_{CC} differs from the the the free-energy functional ΔG_{CC} by the work spent to produce the coupled-cluster apparent charges $\bar{\mathbf{Q}}_N$, corresponding to one-half of the solute-solvent interaction $\bar{\mathbf{Q}}_N(\Lambda, T) \cdot \bar{\mathbf{V}}_N$.

Within the PTE approximation, the T and Λ Eqs. (1.32a) and (1.32b) are independent, and the solution of Eq. (1.32b) is not more necessary to compute the PCM free-energy functional.[12]

References

1. S. Miertuš, E. Scrocco, J. Tomasi, Chem. Phys. **55**, 117 (1981)
2. J. Tomasi, B. Mennucci, R. Cammi, Chem. Rev. **105**, 2999 (2005)
3. J. Tomasi, Theor. Chem. Acc. **103**, 196 (2000)
4. J.L. Pasqual-Ahuir, E. Silla, J. Tomasi, R. Bonaccorsi, J. Comput. Chem. **8**, 778 (1987)
5. J.L. Pasqual-Ahuir, E. Silla, I. Tunon, J. Comput. Chem. **15**, 1147 (1994)
6. E. Cancès, B. Mennucci, J. Tomasi, J. Chem. Phys. **107**, 3092 (1997)
7. R. Cammi, B. Mennucci, K. Ruud, L. Frediani, K.V. Mikkelsen, J. Tomasi, J. Chem. Phys. **117**, 13 (2002)
8. R. Cammi, J. Tomasi, J. Comput. Chem. **16**, 1449 (1995)
9. R. Cammi, J. Chem. Phys. **131**, 164104 (2009)
10. R. Cammi, R. Fukuda, M. Ehara, H. Nakatsuji, J. Chem. Phys. **133**, 024104 (2010)
11. M. Caricato, G. Scalmani, G.W. Trucks, M.J. Frisch, J. Phys. Chem. Lett. **1**, 2369 (2010)
12. M. Caricato, G. Scalmani, M.J. Frisch, J. Chem. Phys. **134**, 244113 (2011)
13. M. Caricato, J. Chem. Phys. **135**, 074113 (2011)
14. F.J. Olivares, del Valle, J. Tomasi, Chem. Phys., **150**, 134 (1991)
15. J. Čížek, Adv. Chem. Phys. **14**, 35 (1969)
16. R.J. Bartlett, in *Modern Electronic Structure Theory*, vol 2, ed. by D.R. Yarkony (World Scientific, Singapore, 1995), p. 1047
17. J. Gauss, in *Encyclopedia of Computational Chemistry*, vol 1, ed. by P.v.R/ Schleyer (Wiley, New York, 1999), p. 617
18. R.J. Bartlett, M. Musiał, Rev. Mod. Phys. **79**, 291 (2007)

---

[12] The explicit equations for the PCM-CC-PTE approximation can be obtained from the PCM-CCSD-PTED equations by neglecting all the explicit terms involving the correlation components of the solvent reaction potential. The resulting PTE-PCM-CCSD equations have the same form of the CCSD equations for an isolated molecule, with the only difference that now the Fock matrix elements and the MO refer to those of the solvated molecules.

Chapter 2
Analytical Derivatives Theory for Molecular Solutes

Abstract This chapter shows how the static properties of the molecular solutes can be expressed as derivatives of the basic energy functional of the PCM model with respect to suitable perturbing fields.

The time-independent properties of the molecular solutes can be expressed as derivatives of the PCM free-energy functional (1.10) with respect to suitable perturbing fields as a consequence of a generalized Helmann-Feynman theorem [1], and analytical expressions and algorithms of these derivatives have been developed for a wide range of perturbations of different nature for the most common quantum chemical levels (see Table 2.1).

As an example of the analytical derivatives of the PCM energy functional, which can be formulated in agreement with a $(2n+1)$ Wigner perturbation rule [1, 2], we will consider the analytical derivatives for the coupled-cluster level up to third order of differentiation.

2.1 The Hellmann-Feynman Theorem for the PCM

The exact eigenfunctions of the effective PCM Hamiltonian (1.12) obey to a generalized Hellmann-Feynman, theorem according to which the first derivative of the free-energy functional G (1.10) with respect to a perturbation parameter λ may be compute as expectation value with the unperturbed wavefunction:

$$\frac{dG}{d\lambda} = < \Psi | H^{[\lambda]} | \Psi > \qquad (2.1)$$

where the operator $H^{[\lambda]}$ is

$$\frac{\partial H}{\partial \lambda} = H^{o,\lambda} + \frac{1}{2}[\langle\Psi|\hat{Q}^{\lambda}|\Psi\rangle \cdot \hat{V} + \langle\Psi|\hat{Q}|\Psi\rangle \cdot \hat{V}^{\lambda}] \qquad (2.2)$$

with the upper-scripts of the operators on the right side of (2.2) denoting differentiation with respect to the parameter λ.

R. Cammi, *Molecular Response Functions for the Polarizable Continuum Model*, Springer Briefs in Electrical and Magnetic Properties of Atoms, Molecules, and Clusters, DOI: 10.1007/978-3-319-00987-2_2, © The Author(s) 2013

Table 2.1 Selected properties of molecular solutes defined as analytical derivatives of the basic PCM free-energy functional The perturbations are **R**, nuclear coordinates, **E** an external electric field, $\mathbf{E}_M$ a Maxwell electric field in the medium (see Appendix), **B** an external magnetic field, $\boldsymbol{\mu}_I$ a magnetic nuclear moment

Property	Derivative
Forces on the nuclei [3]	$\frac{\partial G}{\partial \mathbf{R}}$
Electric dipole moment [3]	$\frac{\partial G}{\partial \mathbf{E}}$
Effective electric dipole moment [4]	$\frac{\partial G}{\partial \mathbf{E}_M}$
Magnetic dipole moment [6]	$\frac{\partial G}{\partial \mathbf{B}}$
Electric polarizability [1]	$\frac{\partial^2 G}{\partial \mathbf{E} \partial \mathbf{E}}$
Nuclear shielding constant [5, 6]	$\frac{\partial^2 G}{\partial \mathbf{B} \partial \boldsymbol{\mu}_I}$
Nuclear Spin–Spin Couplings [7]	$\frac{\partial^2 G}{\partial \boldsymbol{\mu}_J \partial \boldsymbol{\mu}_I}$
Harmonic vibrational frequencies [3, 8–10]	$\frac{\partial^2 G}{\partial \mathbf{R} \partial \mathbf{R}}$
Infrared harmonic absorption intensities [11]	$\frac{\partial^2 G}{\partial \mathbf{R} \partial \mathbf{E}}$
Anharmonic vibrational frequencies [12]	$\frac{\partial^3 G}{\partial \mathbf{R} \partial \mathbf{R} \partial \mathbf{R}}$
Raman intensities (harmonic approximation) [13, 14]	$\frac{\partial^3 G}{\partial \mathbf{R} \partial \mathbf{E} \partial \mathbf{E}}$
Static first electric hyper-polarizability [1]	$\frac{\partial^3 G}{\partial \mathbf{E} \partial \mathbf{E} \partial \mathbf{E}}$

The H-F theorem (2.1) can be derived as follow. The direct differentiation of the free-energy functional (1.10) gives:

$$
\begin{aligned}
\frac{dG}{d\lambda} = & \langle d\Psi/d\lambda | H^o + \tfrac{1}{2}\langle\Psi|\hat{\mathbf{Q}}|\Psi\rangle \cdot \hat{\mathbf{V}}|\Psi\rangle + \langle\Psi|H^o + \tfrac{1}{2}\langle\Psi|\hat{\mathbf{Q}}|\Psi\rangle \cdot \hat{\mathbf{V}}|d\Psi/d\lambda\rangle \\
& + \tfrac{1}{2}[\langle d\Psi/d\lambda|\mathbf{Q}|\Psi\rangle \cdot \langle\Psi|\hat{\mathbf{V}}|\Psi\rangle + \langle\Psi|\mathbf{Q}|\Psi\rangle \cdot \langle d\Psi/d\lambda|\hat{\mathbf{V}}|\Psi\rangle] \\
& + \langle\Psi|H^{o,(\lambda)} + \tfrac{1}{2}[\langle\Psi|\hat{\mathbf{Q}}^\lambda|\Psi\rangle \cdot \hat{\mathbf{V}} + \langle\Psi|\hat{\mathbf{Q}}|\Psi\rangle \cdot \hat{\mathbf{V}}^\lambda]|\Psi\rangle
\end{aligned}
\tag{2.3}
$$

Introducing the identity

$$
\langle d\Psi/d\lambda|\mathbf{Q}|\Psi\rangle \cdot \langle\Psi|\hat{\mathbf{V}}|\Psi\rangle = \langle\Psi|\mathbf{Q}|\Psi\rangle \cdot \langle d\Psi/d\lambda|\hat{\mathbf{V}}|\Psi\rangle
$$

due the symmetry of the kernel of the electrostatic solute-solvent interaction, we can rewrite Eq. (2.3) as

$$
\frac{dG}{d\lambda} = \langle d\Psi/d\lambda | H^o + \langle\Psi|\hat{\mathbf{Q}}|\Psi\rangle \cdot \hat{\mathbf{V}}|\Psi\rangle + c.c. + \langle\Psi|H^{(\lambda)}|\Psi\rangle
\tag{2.4}
$$

In the case of exact eigenfunctions the first two terms of Eq. (2.4) became equal zero, because of the normalization constraint of the wave function, and we obtain the H-F theorem (2.1).

The Helmann-Feynman theorem (2.1) implies that the expectation value of the first-order observable of the molecular solute can be expressed as first derivative of the free energy functional G with respect to a suitable perturbation. If we consider as external perturbation the operator $\lambda\hat{O}$ corresponding to the observable of interest $\hat{O}$ times a scalar factor λ:

$$
H = H^o + \langle\Psi|\hat{\mathbf{Q}}|\Psi\rangle \cdot \hat{\mathbf{V}} + \lambda\hat{O}
\tag{2.5}
$$

from Eqs. (2.1, 2.2) it follows

$$\frac{dG}{d\lambda} = \langle \Psi | \hat{O} | \Psi \rangle \tag{2.6}$$

The H-F theorem (2.1), and its more specific form (2.6), hold also for approximated variational wavefunctions if they are optimized with respect to all the variational parameters (e.g. HF or MCSCF wavefunctions) and if the basis functions used for the expansion of the molecular orbitals are independent on the perturbing parameter λ.

The H-F theorem (2.1) is also involved in formulation of the molecular electronic virial theorem for the PCM model, as shown in Appendix A.2.

2.2 The PCM-CC Analytical Gradients

Let us consider the first derivative of the CM-CC functional ΔG_{CC} (1.20) respect to a perturbation parameter α. As shown in Ref. [15], it can be expressed in the following form:

$$\begin{aligned}
\Delta G_{CC}^{\alpha} = & \langle HF|(1+\Lambda)|e^{-T}[\hat{H(0)}_N^{\alpha} e^T |HF\rangle \\
& + \frac{1}{2}\langle HF|(1+\Lambda)|e^{-T}\hat{Q}_N^{\alpha} e^T |HF\rangle \cdot \bar{\mathbf{V}}_N(\Lambda, T) \\
& + \frac{1}{2}\bar{\mathbf{Q}}_N(\Lambda, T) \cdot \langle HF|(1+\Lambda)|e^{-T}\hat{\mathbf{V}}_N^{\alpha} e^T |HF\rangle
\end{aligned} \tag{2.7}$$

where the upper-script α of the various normal ordered operators denotes the total derivative of their second-quantization form. A key feature of Eq. (2.7) is that it does not contain the first derivative of the T, Λ amplitudes, as a consequence of the stationary of the PCM coupled-cluster energy functional ΔG_{CC} with respect to all the coupled-cluster amplitudes parameters and therefore it represents a generalized Helmann-Feynman theorem for molecular solute described at the coupled-cluster level.

The PCM-CC analytical gradients (2.7) can also rewritten in terms of contraction of differentiated one- and two-electron integrals in the MO basis as

$$\begin{aligned}
\Delta G_{CC}^{\alpha} = & \sum_{ab} f_{ab}^{PCM,\alpha} \gamma_{ab}^{CC-resp} + \sum_{ij} f_{ij}^{PCM,\alpha} \gamma_{ij}^{CC-resp} \\
& + \sum_{ai} f_{ai}^{PCM,\alpha} \gamma_{ai}^{CC-resp} + \sum_{ia} f_{ia}^{PCM,\alpha} \gamma_{ia}^{CC-resp} \\
& + \frac{1}{2}\sum_{pqrs} \mathcal{B}_{pq,rs}^{\alpha} \gamma_{pq}^{CC-resp} \gamma_{rs}^{CC-resp} + \frac{1}{4}\sum_{pqrs} \langle pq||rs\rangle^{\alpha} \Gamma_{rspq}
\end{aligned} \tag{2.8}$$

where $f_{p,q}^{\alpha}$, $\langle pq||rs\rangle^{\alpha}$, $\mathcal{B}_{pq,rs}^{\alpha}$ denote the total first derivatives of the PCM-Fock matrix elements and of the usual two-electron integrals, both in the MO basis;

$\gamma_{pq}^{CC-resp}$ are elements of the one-particle density matrix [15] and Γ_{rspq} are elements of the effective two-particle density matrix Γ [16]. The derivate matrix elements $f_{p,q}^{\alpha}$ may be written in terms of the derivatives of the constituting MO integrals

$$f_{pq}^{PCM,\alpha} = h_{pq}^{\alpha} + j_{pq}^{\alpha} + \sum_{j}^{occ}(\langle pj||qj\rangle^{\alpha} + \mathcal{B}_{jj,pq}^{\alpha}) \tag{2.9}$$

where the derivatives of the solvent integrals j_{pq}^{α} and $\mathcal{B}_{jj,pq}^{\alpha}$ are given by

$$j_{pq}^{\alpha} = \mathbf{v}_{pq}^{\alpha} \cdot \mathbf{q}_{Nuc} + \mathbf{v}_{pq} \cdot \mathbf{q}_{Nuc}^{\alpha} \tag{2.10}$$

$$\mathcal{B}_{pq,rs}^{\alpha} = \mathbf{v}_{\mu\nu}^{\alpha} \cdot \mathbf{q}_{rs} + \mathbf{v}_{\mu\nu} \cdot \mathbf{q}_{rs}^{\alpha} \tag{2.11}$$

where $\mathbf{q}_{p,q}^{\alpha}$ $\mathbf{v}_{p,q}^{\alpha}$ are, respectively, the differentiated apparent charges integrals and the differentiated electrostatic potential integrals [21].

Equation (2.8) is the most general form of analytical gradients of the PCM coupled-cluster functional ΔG_{CC}^{α}, which can be applied to the two alternative forms of the PCM-CC analytical derivatives: the so called non-relaxed MO form [17], which neglects the perturbation α on the MO of the reference determinant, and the so called relaxed MO form, which includes effects of the perturbation of the MO [18]. The use of unrelaxed derivatives may be exploited [17] for the calculation of electrical properties. The approach with orbital relaxation effect is mandatory in all cases where perturbation-dependent basis functions are employed as for example, in the case of geometrical derivatives.

2.3 PCM-CC Analytical Gradients with Relaxed MO

When the MO relaxation is admitted the occupied-virtual block of the derivative of the PCM-Fock matrix, $f_{a,i}^{\alpha}$, vanishes by the stationary condition the Hartree-Fock reference state in the presence of the perturbation, and the gradients ΔG_{CC}^{α} of Eq. (2.8) can be written as:

$$\Delta G_{CC}^{\alpha} = \sum_{ab} f_{ab}^{PCM,\alpha} \gamma_{ab}^{CC-resp} + \sum_{ij} f_{ij}^{PCM,\alpha} \gamma_{ij}^{CC-resp} \tag{2.12}$$

$$+ \frac{1}{2} \sum_{pqrs} (\mathbf{q}_{pq}^{\alpha} \cdot \mathbf{v}_{rs} + \mathbf{q}_{pq} \cdot \mathbf{v}_{rs}^{\alpha}) \gamma_{pq}^{CC-resp} \gamma_{rs}^{CC-resp}$$

$$+ \frac{1}{4} \sum_{pqrs} \langle pq||rs\rangle^{\alpha} \Gamma_{rspq}$$

- The first derivatives of the PCM-Fock matrix of Eq. (2.9) expressed in terms of the derivatives of the MO

$$f_{pq}^{PCM,\alpha} = f_{pq}^{PCM,[\alpha]} + \sum_{r}^{MO} \left(U_{rp}^{\alpha} f_{rq}^{PCM} + U_{rq}^{\alpha} f_{pr}^{PCM} \right) \sum_{r,s} U_{rs}^{\alpha} \delta_{sm} (\langle pr||qs \rangle$$

$$+ \langle ps||qr \rangle + 2B_{pq,rs}) \tag{2.13}$$

where $f_{pq}^{PCM,[\alpha]}$ are the skeleton derivatives of the PCM Fock matrix and U_{mi}^{a} are the derivatives of the MO coefficients $c_{\mu p}$, in MO basis [19], which can be obtained as solutions of the Coupled-Perturbed Hartree-Fock (CPHF) equations properly extended to the PCM model [1, 8].

The skeleton derivatives[1] $f_{pq}^{PCM,[\alpha]}$ are given by

$$f_{pq}^{PCM,[\alpha]} = h_{pq}^{[\alpha]} + j_{pq}^{[\alpha]} + \sum_{j}^{occ} (\langle pj||qj \rangle^{[\alpha]} + \mathcal{B}_{jj,pq}^{[\alpha]}) \tag{2.14}$$

with the skeleton PCM derivative terms $j_{pq}^{[\alpha]}$ and $\mathcal{B}_{pq,kk}^{[\alpha]}$ defined as [15]

$$j_{pq}^{[\alpha]} = \mathbf{v}_{pq}^{[\alpha]} \cdot \mathbf{q}_{Nuc} + \mathbf{v}_{pq} \cdot \mathbf{q}_{Nuc}^{\alpha} \tag{2.15}$$

$$\mathcal{B}_{pq,rs}^{[\alpha]} = \mathbf{v}_{\mu\nu}^{[\alpha]} \cdot \mathbf{q}_{rs} + \mathbf{v}_{\mu\nu} \cdot \mathbf{q}_{rs}^{[\alpha]} \tag{2.16}$$

where $\mathbf{q}_{p,q}^{[\alpha]}$ and $\mathbf{v}_{p,q}^{[\alpha]}$ are, respectively, the skeleton derivatives of the apparent charges integrals and of the electrostatic potential integrals.

- The derivatives of MO integrals $\mathbf{q}_{pq}^{\alpha}$ and $\mathbf{v}_{pq}^{\alpha}$ involved in Eq. (2.12) may be expressed as

$$\mathbf{q}_{pq}^{\alpha} = \mathbf{q}_{pq}^{[\alpha]} + \sum_{r} \left(\mathbf{q}_{rq} U_{rp}^{\alpha*} + \mathbf{q}_{pr} U_{rq}^{\alpha} \right) \tag{2.17}$$

$$\mathbf{v}_{pq}^{\alpha} = \mathbf{v}_{pq}^{[\alpha]} + \sum_{r} \left(\mathbf{v}_{rq} U_{rp}^{\alpha*} + \mathbf{v}_{pr} U_{rq}^{\alpha} \right)$$

- The first derivatives of the two-electron integrals of Eq. (2.12) may be expressed as

$$\langle pq||rs \rangle^{\alpha} = \langle pq||rs \rangle^{[\alpha]} + \sum_{t} \left[\langle tq||rs \rangle U_{tp}^{a} + \langle pt||rs \rangle U_{tq}^{a} \right. \tag{2.18}$$

$$\left. + \langle pq||ts \rangle U_{tr}^{a} + \langle pq||rt \rangle U_{ts}^{a} \right]$$

Then, introducing Eqs. (2.13)–(2.17), and taking into account of the SCF condition for the contractions of the Fock matrix elements involving occupied-occupied and virtual-virtual blocks, the gradients ΔG_{CC}^{α} may be written as

---

[1] The skeleton derivative of integrals on MO basis are the contraction of the differentiated integrals in the AO basis: $x^{[\alpha]pq} = \sum_{\mu\nu} c_{\mu p}^{*} c_{\nu q} x^{[\alpha]\mu\nu}$.

$$\Delta G_{CC}^{\alpha} = \sum_{ab} \gamma_{ab}^{CC-resp} f_{ab}^{PCM,[\alpha]} + \sum_{ij} \gamma_{ij}^{CC-resp} f_{ij}^{PCM,[\alpha]} \tag{2.19}$$

$$+ \frac{1}{2} \sum_{rs} \gamma_{rs}^{CC-resp} \left[\mathbf{q}_{rs}^{[\alpha]} \cdot \bar{\mathbf{V}}_N + \bar{\mathbf{Q}}_N \cdot \mathbf{v}_{rs}^{[\alpha]} \right] + \sum_{pqrs} \Gamma_{pq,rs} \langle pq||rs \rangle^{[\alpha]}$$

$$+ \sum_{pq} I_{pq}' U_{pq}^{\alpha}$$

where I_{pq}' are auxiliary matrix elements (see Tab. VI in Ref. [15]).

Equation (2.19) may be further transformed to avoid the explicit evaluation of the derivative of the MO coefficients U_{ij}^{α} with respect to the specific perturbation α. Using the orthonormality constraint of the perturbed orbitals, we can set the occupied-occupied, virtual-virtual and occupied-virtual CPHF coefficients to $U_{ij}^{\alpha} = -\frac{1}{2} S_{ij}^{[\alpha]}$, $U_{ab}^{\alpha} = -\frac{1}{2} S_{ab}^{[\alpha]}$, $U_{ia}^{\alpha} = -U_{ai} - \frac{1}{2} S_{ia}^{[\alpha]}$. The resulting expression of the PCM-CC gradients is given by:

$$\Delta G_{CC}^{\alpha} = \sum_{ab} \gamma_{ab}^{CC-resp} f_{ab}^{PCM,[\alpha]} + \sum_{ij} \gamma_{ij}^{CC-resp} f_{ij}^{PCM,[\alpha]} \tag{2.20}$$

$$+ \frac{1}{2} \sum_{rs} \gamma_{rs}^{CC-resp} \left[\mathbf{q}_{rs}^{[\alpha]} \cdot \bar{\mathbf{V}}_N + \bar{\mathbf{Q}}_N \cdot \mathbf{v}_{rs}^{[\alpha]} \right] + \sum_{pqrs} \Gamma_{pq,rs} \langle pq||rs \rangle^{[\alpha]}$$

$$+ 2 \sum_{ai} X_{ai} U_{ai}^{\alpha} + \sum_{ij} I_{ij}'' S_{ij}^{[\alpha]} + \sum_{ab} I_{ab}'' S_{ab}^{[\alpha]} - 2 \sum_{ai} I_{ia}'' S_{ai}^{[\alpha]}$$

with $X_{ai} = I_{ai}' - I_{ia}'$ and $I_{pq}'' = I_{qp}'$, if $(p, q) = (a, i)$, otherwise $I_{pq}'' = I_{pq}'$.

2.3.1 The PCM-Z-Vector Method

The block of vacant-occupied CPHF coefficients U_{ai}^{α} of Eq. (2.20), which depend on the specific perturbation α, may be eliminated by using the interchange (Z-vector) method of Handy and Schaefer [20], properly extended to the PCM framework [21]:

$$\sum_{ai} X_{ai} U_{ai}^{\alpha} = \sum_{ai} \gamma_{ai}^{MO-resp} Q_{ai}^{\alpha} \tag{2.21}$$

where $\gamma_{ai}^{MO-resp}$ is the vacant-occupied block of the orbital response one-particle density matrix, independent form the perturbation α, and matrix elements Q_{ai}^{α} are given by

$$Q_{ai}^{\alpha} = f_{ai}^{PCM,[\alpha]} - S_{ai}^{\alpha} f_{ii}^{PCM} - \sum_{kl} S_{kl}^{[\alpha]} \left[\langle al||ik \rangle + \mathscr{B}_{ai,kl} \right]$$

The matrix elements $\gamma_{ai}^{MO-resp}$ are obtained as the solution of a linear system of equations:

$$\sum_{jb} \left[\langle ij||ab \rangle + \langle aj||ib \rangle + 2\mathcal{B}_{ai,bj} \right] \gamma_{bj}^{MO-resp}$$

$$+ \delta_{im}\delta_{ea} \left(f_{ea}^{PCM} - f_{im}^{PCM} \right) \gamma_{ai}^{(MO-resp)} = X_{ai}$$

Then, from Eq. (2.21) the PCM-CC gradients may be expressed in terms of one-particle density matrix independent from the perturbation:

$$\Delta G_{CC}^{\alpha} = \sum_{ab} \gamma_{ab}^{CC-resp} f_{ab}^{PCM,[\alpha]} + \sum_{ij} \gamma_{ij}^{CC-resp} f_{ij}^{PCM,[\alpha]} \qquad (2.22)$$

$$+ \sum_{ij} \gamma_{ai}^{MO-resp} f_{ai}^{PCM,[\alpha]} + \frac{1}{2} \sum_{rs} \gamma_{rs}^{CC-resp} \left[\mathbf{q}_{rs}^{[\alpha]} \cdot \bar{\mathbf{V}}_N + \bar{\mathbf{Q}}_N \cdot \mathbf{v}_{rs}^{[\alpha]} \right]$$

$$+ \sum_{pqrs} \Gamma_{pq,rs} \langle pq||rs \rangle^{[\alpha]} + \sum_{ij} \tilde{I}_{ij} S_{ij}^{[\alpha]} + \sum_{ab} I''_{ab} S_{ab}^{[\alpha]} - 2 \sum_{ai} \tilde{I}_{ia} S_{ai}^{[\alpha]}$$

where $\tilde{I}_{ia} = I''_{ia} - \gamma_{ai}^{MO-resp} f_{ii}^{PCM}$, and $\tilde{I}_{ij}$ are given by

$$\tilde{I}_{ij} = I''_{ij} - \sum_{em} \gamma_{em}^{MO-resp} \left(\langle ei||mj \rangle + \langle im||je \rangle + 2\mathcal{B}_{em,ij} \right)$$

The expression of the PCM-CC gradients can be easily reverted from the MO to the AO representation of the matrices elements:

$$\Delta G_{CC}^{\alpha} = \sum_{\mu\nu} \gamma_{\mu\nu}^{CC-MO} \left(h_{\mu\nu}^{\alpha} + j_{\mu\nu}^{\alpha} \right) + \sum_{\mu\nu} I'_{\mu\nu} S_{\mu\nu}^{\alpha} \qquad (2.23)$$

$$+ \sum_{\mu\nu\rho\sigma} (\gamma_{\mu\nu}^{CC-MO} P_{\sigma\rho}^{HF} + \frac{1}{2} \gamma_{\mu\nu}^{CC-resp} \gamma_{\sigma\rho}^{CC-resp}) \mathcal{B}_{\mu\nu\sigma\rho}^{\alpha}$$

$$+ \sum_{\mu\nu\sigma\rho} \Gamma'_{\mu\nu,\sigma\rho} \langle \mu\sigma||\rho\sigma \rangle^{\alpha}$$

where $\gamma_{\mu\nu}^{CC-MO} = \gamma_{\mu\nu}^{CC-resp} + \gamma_{\mu\nu}^{MO-resp}$ $\Gamma'_{\mu\nu,\sigma\rho} = \Gamma_{\mu\nu,\sigma\rho} + \gamma_{\mu\nu}^{CC-MO} P_{\sigma\rho}^{HF}$. A similar expression can be derived for the analytical gradients of the PCM-CC-PTE approximation [15].

2.4 PCM Analytical Derivatives with Unrelaxed MO: The 2n+1 Rule

We now consider the first and higher orders of the PCM-CC free-energy functional (1.20) within the approximation which neglects the effect of the perturbations on the MO of the Hartree-Fock reference state $|HF\rangle$. The derivatives will be with respect to the amplitude parameters $\alpha, \beta, \ldots$, associated to the perturbing operators $X, Y, \ldots$, of the effective Hamiltonian effective Hamiltonian of the molecular solute:

$$H_N = H(0)_N + \bar{\mathbf{Q}}_N(T, \Lambda) \cdot \mathbf{V}_N + \alpha X_N + \beta Y_N + \cdots \tag{2.24}$$

where $X_N, Y_N, \ldots$ denote the normal ordered form of the perturbing operators ($X_N = X - \langle HF|X|HF\rangle$, $Y_N = Y - \langle HF|X|HF\rangle$).

- The first derivative ΔG_{CC}^{α}, with respect to the perturbation α of Eq. (2.24) is given by:

$$\Delta G_{CC}^{\alpha} = \langle HF|(1 + \Lambda)e^{-T} X_N e^{T}|HF\rangle \tag{2.25}$$

Equation (2.25) follows directly from Eq. (2.7), and represents a form of the generalized Hellmann-Feynman theorem (2.1) for the Coupled-cluster method.

- The second derivative $\partial^2 \Delta G_{CC}/\partial\alpha\partial\beta = \Delta G_{CC}^{\alpha\beta}$, can be obtained by differentiation of the analytical gradients of Eq. (2.25) with respect to a second perturbation β:

$$\begin{aligned}
\Delta G_{CC}^{\beta\alpha} = & \langle HF|\tfrac{\partial\Lambda}{\partial\beta}e^{-T} X e^{T}|HF\rangle \\
& + \langle HF|(1 + \Lambda)[e^{-T} X e^{T}, \tfrac{\partial T}{\partial\beta}]|HF\rangle
\end{aligned} \tag{2.26}$$

where T^{β} and Λ^{β} are the first order corrections of the T and Λ cluster amplitudes, respectively, with respect to the perturbation amplitude β.

The first order amplitudes T^{β} can be determined by the perturbative expansion of the T amplitude Eq. (1.24) limited to the first order in the perturbation:

$$0 = \langle HF|\tau_p^{\dagger}e^{-T} \tilde{Y} e^{T}|HF\rangle + \langle HF|\tau_p^{\dagger}[e^{-T} H_N e^{T}, \tfrac{\partial T}{\partial\beta}]|HF\rangle \tag{2.27}$$

The first order amplitudes Λ^{β} can be determined in a similar way from the Λ Eq. (1.28)

$$\begin{aligned}
0 = & \langle HF|(1 + \Lambda)[e^{-T} \tilde{Y} e^{T}, \tau_p]|HF\rangle \\
& + \langle HF|(1 + \Lambda)[[e^{-T} H_N e^{T}, \tfrac{\partial T}{\partial\beta}], \tau_p]|HF\rangle \\
& + \langle HF|\tfrac{\partial\Lambda}{\partial\beta}e^{-T} [H_N, \tau_p]e^{T}|HF\rangle
\end{aligned} \tag{2.28}$$

where $\tilde{Y}$ is an effective perturbing operator:

$$\tilde{Y}_N = Y_N + \bar{Q}_N^{\beta} \cdot V_N \tag{2.29}$$

being Y the operator representing the external perturbation having amplitude β, and $\bar{Q}_N^Y$ is given by

$$\bar{Q}_N^{\beta} = \langle HF|(1+\Lambda)e^{-T}[Q_N, \tfrac{\partial T}{\partial \beta}]e^T|HF\rangle + \langle HF|\tfrac{\partial \Lambda}{\partial \beta}e^{-T}Q_N e^T|HF\rangle \tag{2.30}$$

An alternative expression of the second derivative $\Delta G_{CC}^{\alpha\beta}$ can be obtained by differentiating Eq. (1.20) and eliminating the second-order terms of the derivatives of the T, Λ amplitudes, as a consequence of the stationary of the unperturbed free-energy functional, in agreement with the (2n+1) rule:

$$\begin{aligned}
\Delta G_{CC}^{\beta\alpha} = & \; P(XY)\langle HF|(1+\Lambda)e^{-T}|[X, \tfrac{\partial T}{\partial \beta}]e^T|HF\rangle \\
& + \langle HF|(1+\Lambda)e^{-T}|[[H_N, \tfrac{\partial T}{\partial \alpha}], \tfrac{\partial T}{\partial \beta}]e^T|HF\rangle \tfrac{1}{2}P(XY) \\
& \langle HF|(1+\Lambda)e^{-T}[Q_N, \tfrac{\partial T}{\partial \alpha}]e^T|HF\rangle \\
& \cdot \langle HF|(1+\Lambda)e^{-T}[V_N, \tfrac{\partial T}{\partial \beta}]e^T|HF\rangle \tfrac{1}{2}P(XY) \\
& \langle HF|\tfrac{\partial \Lambda}{\partial \alpha}Q_N e^T|HF\rangle \cdot \langle HF|\tfrac{\partial \Lambda}{\partial \beta}V_N e^T|HF\rangle
\end{aligned} \tag{2.31}$$

where $P(x, y)$ is a permutation operator between x and y.

- The third derivatives $\partial^3 \Delta G_{CC}/\partial\alpha\partial\beta\partial\gamma = \Delta G_{CC}^{\alpha\beta\gamma}$ can be obtained differentiating three times Eq. (1.20) and eliminating higher-order terms according to the (2n+1) rule[2]:

$$\begin{aligned}
\Delta G_{CC}^{\alpha\beta\gamma} = & \; P^3(XYZ)\langle HF|(1+\Lambda)e^{-T}|[[X, \tfrac{\partial T}{\partial \beta}, \tfrac{\partial T}{\partial \gamma}]e^T|HF\rangle \\
& + P^6(XYZ)\langle HF|\tfrac{\partial \Lambda}{\partial \alpha}e^{-T}[Y, \tfrac{\partial T}{\partial \gamma}]e^T|HF\rangle \\
& + \langle HF|(1+\Lambda)e^{-T}[[[H_N, \tfrac{\partial T}{\partial \alpha}], \tfrac{\partial T}{\partial \beta}], \tfrac{\partial T}{\partial \gamma}]e^T|HF\rangle \\
& + P^3(XYZ)\langle HF|\tfrac{\partial \Lambda}{\partial \alpha}e^{-T}[[H_N, \tfrac{\partial T}{\partial \beta}], \tfrac{\partial T}{\partial \gamma}]e^T|HF\rangle \\
& \tfrac{1}{2}P^3(XYZ)\left(\tilde{Q}_N^{\alpha\beta} \cdot \bar{V}_N^{\gamma} + \bar{Q}_N^{\alpha} \cdot \tilde{V}_N^{\beta\gamma}\right)
\end{aligned} \tag{2.32}$$

where $\bar{V}_N^{\beta}$ is defined by an expression similar to Eq. (2.30), while the explicit solvent factors $\tilde{Q}_N^{\alpha}$, and $\tilde{V}_N^{\beta\gamma}$ are defined, respectively, as:

$$\begin{aligned}
\tilde{Q}_N^{\alpha\beta} = & \; \langle HF|(1+\Lambda)e^{-T}[[Q_N, \tfrac{\partial T}{\partial \alpha}, \tfrac{\partial T}{\partial \beta}]e^T|HF\rangle \\
& + P(XY)\langle HF|\tfrac{\partial \Lambda}{\partial \alpha}e^{-T}[Q_N, \tfrac{\partial T}{\partial \beta}]e^T|HF\rangle
\end{aligned}$$

---

[2] Third- and second-order derivatives of the T, Λ amplitudes are eliminated, respectively, by the zero- and first order stationarity of the free-energy functional ΔG_{CC}.

$$\tilde{\mathbf{V}}_N^{\alpha\beta} = \langle HF|(1+\Lambda)e^{-T}[[\mathbf{V}_N, \tfrac{\partial T}{\partial \alpha}, \tfrac{\partial T}{\partial \beta}]e^T|HF\rangle$$
$$+ P(XY)\langle HF|\tfrac{\partial \Lambda}{\partial \alpha}e^{-T}[\mathbf{V}_N, \tfrac{\partial T}{\partial \beta}]e^T|HF\rangle$$

The symbols $P^3(xyz)$ and $P^6(xyz)$ denotes, respectively the cyclic permutation operator and the full permutation operator of their arguments.

References

1. R. Cammi, M. Cossi, J. Tomasi, J. Chem. Phys. **104**, 4611 (1996)
2. J.G. Angyan, J. Math. Chem. **104**, 4611 (2000)
3. R. Cammi, J. Tomasi, J. Chem. Phys. **100**, 7495 (1994)
4. R. Cammi, B. Mennucci, J. Tomasi, J. Phys. Chem. A **104**, 4690 (2000)
5. R. Cammi, J. Chem. Phys. **109**, 3185 (1998)
6. R. Cammi, B. Mennucci, J. Tomasi, J. Chem. Phys. **110**, 7627 (1999)
7. K. Ruud, L. Frediani, R. Cammi, B. Mennucci, J. Mol. Sci. **4**, 119 (2003)
8. B. Mennucci, R. Cammi, J. Tomasi, J. Chem. Phys. **110**, 6858 (1999)
9. M. Cossi, G. Scalmani, N. Rega, V. Barone, J. Chem. Phys. **117**, 43 (2002)
10. R. Cammi, B. Mennucci, C. Pomelli, C. Cappelli, S. Corni, L. Frediani, G.W. Trucks, M.J. Frisch, Theor. Chem. Acc. **111**, 66 (2004)
11. R. Cammi, C. Cappelli, S. Corni, J. Tomasi, J. Phys. Chem. A **104**, 9874 (2000)
12. C. Cappelli, S. Monti, G. Scalmani, V. Barone, J. Chem. Theory Comput. **6**, 1660 (2010)
13. S. Corni, C. Cappelli, R. Cammi, J. Tomasi, J. Phys. Chem. A **105**, 8310 (2001)
14. G. Scalmani, M.J. Frisch, J. Chem. Phys. **132**, 114110 (2010)
15. R. Cammi, J. Chem. Phys. **131**, 164104 (2009)
16. R. Bartlett, M. Musiał, Rev. Mod. Phys. **79**, 291 (2007)
17. H. Koch, P. Jrgensen, J. Chem. Phys. **93**, 3333 (1990)
18. J. Gauss, in *Encyclopedia of Computational Chemistry*, vol. I, ed. by P.V.R Schleyer (Wiley, New York, 1999), p. 617
19. Y. Yamaguchi, J.D. Goddard, Y. Osamura, H.F. Schaefer III, *A New Dimension in Quantum Chemistry: Analytical Derivative Methods in Ab-initio Molecular Electronic Structure Theory* (Oxford University Press, New York, 1999)
20. N.C. Handy, H.F. Schaefer III, J. Chem. Phys. **81**, 5031 (1981)
21. R. Cammi, B. Mennucci, J. Tomasi, J. Phys. Chem. A. **103**, 9100 (1999)
22. M. Caricato, G. Scalmani, G.W. Trucks, M.J. Frisch, J. Phys. Chem. Lett. **1**, 2369 (2010)
23. M. Caricato, G. Scalmani, M.J. Frisch, J. Chem. Phys. **134**, 244113 (2011)
24. M. Caricato, J. Chem. Phys. **135**, 074113 (2011)
25. R. Cammi, R. Fukuda, M. Ehara, H. Nakatsuji, J. Chem. Phys. **133**, 024104 (2010)

Chapter 3
General Response Theory for the Polarizable Continuum Model

Abstract This chapter presents the general aspects of the response theory for molecular solutes in the presence of time-dependent perturbing fields: (i) the non-equilibrium solvation, (ii) the variational formulation of the time-dependent non-linear QM problem, and (iii) the connection of the molecular response functions with their macroscopic counterparts. The linear and quadratic molecular response functions are described at the coupled-cluster level.

The response functions theory the PCM method [1] is an extension of the response theory for molecules in the gas phase [2, 3]. This latter is based on a variational-perturbation approach for the description of the variations of the electronic wave function and of the changes of the observables properties at the various orders of perturbation with respect to the perturbing fields, and no restrictions are posed on the nature of the observables and on the nature of the perturbing fields, and the theory gives also access to a direct determination of the transition properties (i.e. transition energies and transition probabilities) associated with transitions between the stationary states of the molecular systems. The PCM response theory adds to this framework several new elements.

The new elements are required to describe new interaction phenomena that involve, in an entangled way, the molecular system, the medium (i.e. the solvent) and the external time-dependent perturbing fields. The couplings has two main effects. The first effect is to add a non-linearity in the time-dependent QM problem. More specifically, the time-dependent QM problem requires a proper extension of the time-dependent variation principle. The second effect is to induce a time-dependence in the solute-solvent responsive interaction, which must be described in a non-equilibrium solvation scheme.[1] The PCM response theory face also the complex problem of the connection of the response functions of the molecular solutes

---

[1] The concept of non-equilibrium solvation has been introduced to describe the solvent polarization in processes involving dynamic, or sudden, variations of solute charge distribution of the solute, and it takes into account that during the time-scale of a fast event not all the degrees of freedoms of the solvent molecules (nuclear, translational, rotational, vibrational; electronic) are able to respond to the variations of the charge distribution of the solute (see Appendix).

R. Cammi, *Molecular Response Functions for the Polarizable Continuum Model*,
Springer Briefs in Electrical and Magnetic Properties of Atoms, Molecules, and Clusters,
DOI: 10.1007/978-3-319-00987-2_3, © The Author(s) 2013

with the corresponding macroscopic counterpart (i.e. the macroscopic susceptibilities) measured in the experiments, as the molecular solutes are locally subjected to perturbing fields which are different to that measured by the experimenter.

The PCM response function theory has been developed for linear and non-linear response functions and at several QM levels, including the variational wavefunctions methods (SCF,MCSCF) and the Density Functional Theory [4–18] and more recently at the Coupled-Cluster level [19, 20].

3.1 The Variational Time-Dependent Theory for the Polarizable Continuum Model

Let us consider the time-dependent non-linear Schrödinger equation (TDNLSE) for a molecular solutes

$$i\frac{\partial}{\partial t}|\bar{\Psi}(t)> = H|\bar{\Psi}(t)> \tag{3.1}$$

where H is the effective time-dependent Hamiltonian for the solute:

$$H = H^0 + \bar{\mathbf{Q}}(\Psi; t) \cdot \mathbf{V} + V'(t) \tag{3.2}$$

where H^0 is the Hamiltonian of the isolated molecule, $\bar{\mathbf{Q}}(\Psi; t) \cdot \mathbf{V}$ is the potential energy term representing the solute-solvent electrostatic interaction, and $V'(t)$ is a generic time-dependent external perturbation.[2]

In Eq. (3.2), $\mathbf{Q}(\Psi; t)$ represents the polarization charges induced by the solute on the boundary of the cavity hosting the solute within the dielectric medium representing the solvent, the dot represents a vectorial inner product, and $\hat{V}$ is a vectorial operator representing the electrostatic potential of the solute at the boundary cavity. The solvent polarization charges collected in $\mathbf{Q}(\Psi; t)$ depend parametrically on time, as they are determined to by the first-order density matrix $\Psi^*(t)\Psi(t)$ of the molecular solute; the most general form of $\bar{\mathbf{Q}}(\Psi; t)$ must take into account of the non-equilibrium solvation effects related to the intrinsic dynamics of solvent polarization.

In the presence of a generic time-dependent perturbation $V(t)$, the time-dependent polarization charges $\mathbf{Q}(\Psi; t)$ may be formally defined as:

$$\bar{\mathbf{Q}}(\Psi; t) = \int_{-\infty}^{t} < \Psi(t')|\mathbf{Q}(t - t')|\Psi(t') > dt' \tag{3.3}$$

where $\mathbf{Q}(t - t')$ is an apparent charge operator non-local in time, describing a general non-equilibrium solvation regime where, the description of polarization of the medium cannot be given in terms of a single value of the dielectric permittivity of the medium (as in the case of the equilibrium solvation), but it requires in principle

---

[2] We assume that $V'(t)$ is applied adiabatically so that it vanishes at $t = -\infty$.

knowledge of the whole spectrum of the frequency-dependent dielectric permittivity $\varepsilon(\omega)$ [21].

3.1.1 The Time-Dependent Quasi-Free-Energy and its Variational Principle

It has been shown [22] that the time-dependent non- linear Schrödinger Eq. (3.1) can be obtained from the Hamilton principle when a suitable QM Lagrangian density is defined (see Appendix A.1), and that this Lagrangian density implies an extension of the time-dependent Frenkel's variational principle [23].

Let us consider the time-dependent wave function $\bar{\Psi}(t) >$ expressed in the phase isolated form

$$|\bar{\Psi}(t) >= e^{-iF(t)}|\Psi(t) > \tag{3.4}$$

where $F(t)$ is a function of time, and $|\Psi(t) >$ is the so-called regular wave function, which depends only parametrically on time and which reduces in the unperturbed limit (i.e. $t = -\infty$) to the time-independent wave function describing a stationary state of the molecular solute.

The phase $F(t)$ is determined by substituting Eq. (3.4) into the TDNLSE (3.1):

$$\dot{F}(t) =< \Psi(t)|H - i\frac{\partial}{\partial t}|\Psi(t) >$$

The regular wave function $|\Psi(t) >$ satisfies a generalized time-dependent Frenkel variational principles [24] for an arbitrary variation $\delta\Psi$

$$\delta\mathscr{G}(t) + i\frac{\partial}{\partial t} < \Psi|\delta\Psi >= 0 \tag{3.5}$$

where $\mathscr{G}(t)$ is the time-dependent quasi-free energy functional [19]

$$\mathscr{G}(t) =< \Psi(t)|H^o + \frac{1}{2}\bar{\mathbf{Q}}(\Psi; t) \cdot \mathbf{V} - i\frac{\partial}{\partial t}|\Psi(t) > \tag{3.6}$$

that in the time-independent (i.e. unperturbed) limit reduces to the time-independent PCM free energy functional $G =< \Psi|H^o + \frac{1}{2}\bar{\mathbf{Q}}(\Psi) \cdot \mathbf{V}|\Psi >$.

Let us now consider the case of a periodic time-dependent perturbation:

$$V(t + T; \varepsilon) = V(t; \varepsilon) \tag{3.7}$$

with period T, frequency $\omega = \frac{2\pi}{T}$, and perturbation strength ε. In this case, the phase-isolated form of the wavefunction acquires the same periodicity T[3]:

$$|\Psi(t+T)> = |\Psi(t)> \tag{3.8}$$

and satisfies the stationarity condition:

$$\delta\{\mathscr{G}(t)\}_T = \delta\{< \Psi(t)|H^o + \frac{1}{2}\bar{\mathbf{Q}}(\Psi;t)\cdot\mathbf{V} - i\frac{\partial}{\partial t}|\Psi(t)\}_T = 0 \tag{3.9}$$

where $\{\mathscr{G}(t)\}_T$ is the time-averaged[4] quasi-free-energy functional

$$\{G(t)\}_T = \frac{1}{T}\int_{-T/2}^{T/2} < \Psi(t)|H^o + \frac{1}{2}\bar{\mathbf{Q}}(\Psi;t)\cdot\mathbf{V} - i\frac{\partial}{\partial t}|\Psi(t) > dt \tag{3.10}$$

From the variational condition (3.9), $\{\mathscr{G}(t)\}_T$ satisfies a generalization of the time-dependent Hellmann–Feynman theorem, and if we consider in the Hamiltonian (3.2) an external perturbation $\varepsilon V(\omega)$ with amplitude ε and periodicity $T = 2\pi/\omega$, we obtain

$$\begin{aligned}\frac{d\{G(t)\}_T}{d\varepsilon} &= \left\{\langle\Psi(t)|\frac{\partial H}{\partial\varepsilon}|\Psi(t)\rangle\right\}_T \\ &= \{\langle\Psi(t)|V(\omega)|\Psi(t)\rangle\}_T\end{aligned} \tag{3.11}$$

where $\{\langle\Psi(t)|V(\omega)|\Psi(t)\rangle\}$ is the time-average of the expectation value of the perturbing operator $V(\omega)$.

The time-dependent stationary conditions (3.9) and the corresponding Hellmann–Feynman theorem (3.11) are the basic equations for the determination of the response functions of the molecular solutes in the presence of periodic external perturbations.

3.1.2 The Response Functions of Molecular Solutes

Let us consider the case of a time-dependent perturbation $V(t)$ having several periodic components:

---

[3] The time-dependent wavefunction (3.8) are also known as Floquet states [24–28], a particular set of solutions of the time-dependent Schrödinger equation for systems under the influence of an external time-dependent periodic perturbation.

[4] The time-average over a period T for a general time-dependent function $g(t)$ is defined as

$$\{g(t)\}_T = \frac{1}{T}\int_{-T/2}^{T/2} g(t)dt$$

$$V(t) = \sum_{j=-K}^{K} \sum_{x} e^{(-i\omega_j t)} \boldsymbol{\varepsilon}_X(\boldsymbol{\omega}_j) X \qquad (3.12)$$

where X is a perturbation operator, $\varepsilon_X(\omega_i)$ is the perturbation amplitude, and ω_i is the corresponding frequency.[5] In the case of a perturbing agent given by an electromagnetic field in the medium, (i.e. the macroscopic Maxwell field), the X operator corresponds to an effective electric dipole operator which takes into account the effect of the cavity boundary on the Maxwell field (see Sect. 3.3 below and Appendix A.4) [33]. For the sake of simplicity, in the following we assume the X operator to be the interaction operator of the molecular solute with a perturbing agent in the medium, unmodified by the cavity boundary.

The response functions are the coefficients of a Fourier-perturbation expansion of the time-dependent expectation values of a generic observable X, $< X(t) > = < \Psi(t)|X|\Psi(t) >$ of the molecular solute in order of the perturbations $V(t)$ [2]:

$$\begin{aligned} < \Psi(t)|X|\Psi(t) > &= < X >_0 + \sum_{j,y} e^{(-i\omega_i t)} << X; Y >>_{\omega_j} \varepsilon_y(\omega_i) \\ &+ \tfrac{1}{2} \sum_{i,j,y,z} e^{(-i(\omega_i + \omega_j)t)} << X; Y, Z >>_{\omega_y + \omega_z} \varepsilon_y(\omega_i)\varepsilon_z(\omega_j) + \cdots \end{aligned} \qquad (3.13)$$

where $<< X; Y >>_{\omega_Y}$ represents the linear-response function describing the contribution of $X(t)$ of first-order in the perturbation Y with frequency ω_Y, $<< X; Y, Z >>_{\omega_Y, \omega_Z}$ denotes the quadratic-response function describing the contribution of $X(t)$ quadratic in the perturbations Y, Z with frequencies ω_Y, ω_Z, and in a similar way are defined the higer-order response functions $<< X; Y, Z, .. >>_{\omega_Y, \omega_Z, ...}$ (see Table 3.1).

Within the variational time-dependent approach of Sect. 3.1.1, the molecular response functions (3.13) are determined by expanding the time-dependent wavefunction $|\Psi(t) >$ and the time-averaged free-energy functional (3.10) in orders of the perturbation, and by imposing that the variational condition (3.9) is satisfied at the various order. The response functions are then identified by means of the Hellmann–Feynman theorem (3.11), as terms of the expansion of the quasi-free-energy.

3.2 The Coupled-Cluster Response Functions Theory

The time-dependent coupled-cluster wavefunction may be expressed in the phase-isolated form as [29]:

$$|CC(t) >= e^{-iF(t)} e^{T(t)} |HF > \qquad (3.14)$$

---

[5] Being the perturbing operator $V(t)$ Hermitian, we have that: $X^\dagger = X$, $\omega_{-j} = -\omega_j$, $\varepsilon(\omega_j)^* = \varepsilon(\omega_{-j})$.

Table 3.1 Selected response functions $<< X; Y, Z, .. >>_{\omega_Y, \omega_Z, ...}$ of molecular solutes. The operators are μ_α, Cartesian components of the electric dipole operator, $\Theta_{\alpha, \beta}$, element of the electric quadrupolar tensor, m_α Cartesian component of the magnetic dipole operator

Response function	Property
$<< \mu_\alpha; \mu_\beta >>_\omega$ [4]	Dipole polarizability $\alpha_{\alpha, \beta}(-\omega; \omega)$
$i << m_\alpha; \mu_\beta >>_\omega$ [7]	Optical rotation tensor $G'_{\alpha, \beta}(\omega)$
$<< \mu_\alpha; \Theta_\beta >>_\omega$ [15]	Quadrupolar polarizability tensor $\alpha_{\alpha, \beta}(-\omega; \omega)$
$<< \mu_\alpha; \mu_\beta, \mu_\gamma >>_\omega$ [4]	First polarizability $\beta_{\alpha, \beta, \gamma}(-\omega_\alpha; \omega_\beta, \omega_\gamma)$
$<< \mu_\alpha; \mu_\beta, \mu_\gamma, \mu_\delta >>_{\omega_\alpha, \omega_\beta, \omega_\delta}$ [4]	Second polarizability $\beta_{\alpha, \beta, \gamma}(-\omega_\alpha; \omega_\beta, \omega_\gamma)$
$- << \mu_\alpha; \mu_\beta, m_\gamma, m_\delta >>_{\omega, 0, 0}$	
$+ << \mu_\alpha; \mu_\beta, m_\gamma, m_\delta >>_{\omega, 0, 0}$ [11]	Frequency dependent mixed electric magnetic hyper-magnetizability $\eta_{\alpha, \beta, \gamma, \delta}(-\omega; \omega, 0, 0)$

where the reference state $|HF>$ is the fixed, time-independent Hartree–Fock ground state of the solvated molecules[6]; the phase factor $F(t)$ is a function of time and $T(t)$ is the time-dependent cluster excitation operator.

The coupled-cluster phase factor $F(t)$ is then determined as:

$$\dot{F}(t) = < HF|(1 + \Lambda)e^{-T}\left(H(0)_N + \bar{\mathbf{Q}}_N(t) \cdot \mathbf{V}_N + V(t) - i\frac{\partial}{\partial t}\right)e^T|HF>$$

where $\Lambda(t)$ is the time-dependent coupled-cluster de-excitation operator, and the time-dependent apparent charges $\bar{\mathbf{Q}}_N(t)$ are formally given as

$$\bar{\mathbf{Q}}_N(t) = \int_\infty^t < HF|(1 + \Lambda(t))e^{-T(t)}\mathbf{Q}_N(t, t')e^{T(t)}|HF> dt' \qquad (3.15)$$

where $\mathbf{Q}_N$ is an apparent charge operator non-local with respect to the time (see Eq. (3.3), and Λ is a time-dependent de-excitation operator.

The time-dependent coupled-cluster $T(t)$ and $\Lambda(t)$ amplitudes are obtained from the variational criterion (see Eq. 3.9) applied to the following quasi-free-energy functional

$$\{\Delta G_{CC}(t)\}_T = \frac{1}{T}\int_{-T/2}^{T/2} < HF|(1 + \Lambda(t))e^{-T(t)}\left[H(0)_N + \frac{1}{2}\bar{\mathbf{Q}}_N(t) \cdot \mathbf{V}_N + V(t) - i\frac{\partial}{\partial t}\right]e^T(t)|HF> dt \qquad (3.16)$$

where the time integration is over a common multiple of periods, T, such that $V(t + T) = V(t)$.

The corresponding variational condition may be written as

$$\delta\{\Delta G_{CC}(t)\}_T = 0 \qquad (3.17)$$

---

[6] We here consider the Molecular Orbital (MO) "unrelaxed" approach in which the reference state does not depend on the perturbation $V(t)$.

Equation (3.17) is then solved by a variational-perturbative procedure, where the coupled-cluster T and Λ amplitudes are expanded with respect to the components of the perturbation $V(t)$:

$$T(t) = T^{(0)} + T^{(1)} + T^{(2)}... \tag{3.18a}$$

$$\Lambda(t) = \Lambda^{(0)} + \Lambda^{(1)} + \Lambda^{(2)}... \tag{3.18b}$$

with:

$$\begin{aligned} T^{(1)} &= \sum_i exp(-i\omega_i t) \sum_x \varepsilon_X(\omega_i) T^X(\omega_i) \\ T^{(1)} &= \tfrac{1}{2} \sum_{i,j} exp(-i(\omega_i + \omega_j)t) \sum_{X,Y} \varepsilon_X(\omega_i)\varepsilon_Y(\omega_j) T^{XY}(\omega_i, \omega_j)... \end{aligned} \tag{3.19a}$$

$$\begin{aligned} \Lambda^{(1)} &= \Lambda^{(0)} + \sum_i exp(-i\omega_i t) \sum_x \varepsilon_X(\omega_i) \Lambda^X(\omega_i) \\ \Lambda^{(2)} &= \tfrac{1}{2} \sum_{i,j} exp(-i(\omega_i + \omega_j)t) \sum_{X,Y} \varepsilon_X(\omega_i)\varepsilon_Y(\omega_j) \Lambda^{XY}(\omega_i, \omega_j)... \end{aligned} \tag{3.19b}$$

The expansions (3.19a), (3.19b), lead to a corresponding expansion of the quasi-free energy functional (3.16):

$$\{\Delta G(t)_{CC}\}_T = \left\{\Delta G_{CC}^{(0)}\right\}_T + \left\{\Delta G_{CC}^{(1)}\right\}_T + \left\{\Delta G_{CC}^{(2)}\right\}_T + ... \tag{3.20}$$

with the n-th order term $\left\{\Delta G_{CC}^{(n)}\right\}_T$ given by

$$\left\{\Delta G_{CC}^{(n)}\right\}_T = \frac{1}{n!} \sum_{i,j,...} \sum_{x,y,z} \varepsilon_X(\omega_i)\varepsilon_Y(\omega_j) \cdots \times G^{XY\cdots}(\omega_i, \omega_j, \cdots) \tag{3.21}$$

where $G^{XY\cdots}(\omega_i, \omega_j, \cdots)$ are defined as

$$G^{XY\cdots}(\omega_i, \omega_j, \cdots) = \frac{d^n \left\{\Delta G^{(n)}\right\}_T}{d\varepsilon_i d\varepsilon_j \ldots} \tag{3.22}$$

with $\omega_i + \omega_j + \cdots = 0$.[7]

Then, introducing the perturbation expansion of Eq. (3.20) into the time-average variational criterion (3.17), and separating by orders, we arrive to a set of variational criteria on the Fourier components $G^{XY\cdots}(\omega_i, \omega_j, \cdots)$ of Eq. (3.22)

---

[7] The expansions (3.18a and 3.18b) imply an expansion in time-dependent polarization charges of (3.15):

$$\bar{Q}_N(t) = \bar{Q}_N(t)^{(0)} + \bar{Q}_N(t)^{(1)} + \cdots$$

where $\bar{Q}_N(t)^{(n)}$ are terms of n-th order in the perturbation.

$$\delta G^{XY\cdots}(\omega_i, \omega_j, \cdots) = 0 \qquad (3.23)$$

that determine the Fourier coefficients $T^{XY\cdots}(\omega_i, \omega_j, \cdots)$ and $\Lambda^{XY\cdots}(\omega_i, \omega_j, \cdots)$ for the T and Λ amplitudes (3.19a) and (3.19b).[8]

When the variational conditions (3.23) are satisfied, the response functions of the molecular solutes can be recognize as the stationary Fourier components $G^{XY\cdots}(\omega_i, \omega_j, \cdots)$:

$$
\begin{aligned}
<< X; Y >>_{\omega_i} &= \tilde{G}^{XY}(\omega_i, \omega_j) \\
<< X; Y, Z >>_{\omega_i} &= \tilde{G}^{XYZ}(\omega_i, \omega_j, \omega_k) \\
\vdots \qquad &= \vdots \\
<< X; Y, Z, \cdots >>_{\omega_i} &= \tilde{G}^{XYZ\cdots}(\omega_i, \omega_j, \dots)
\end{aligned}
\qquad (3.24)
$$

where the tilde peaks denotes stationarity. The response functions (3.24) satisfy a generalized $(2n + 1)$ rule with respect to the perturbed T and Λ amplitudes.

3.2.1 Coupled-Cluster Linear Response Equations

Let us consider the first order equations for the coupled-cluster Fourier amplitudes $\Lambda^X(\pm\omega)\, T^X(\pm\omega)$. They are determined from the stationary of the second order time-averaged quasi-free energy $G^{XY}(-\omega, \omega)$:

$$
\begin{aligned}
G^{XY}(-\omega, \omega) =\ & P(XY) < HF|(1 + \Lambda)e^{-T}|[X, T^Y(\omega)]e^T|HF > \\
& + < HF|(1 + \Lambda)e^{-T}[[H_N, T^X(-\omega)], T^Y(\omega)]e^T|HF > \\
& - \omega P(XY) < HF|\Lambda^X(-\omega)T^Y(\omega)|HF > \\
& + P(XY) < HF|\Lambda^X(-\omega)e^{-T}[Y + [H_N, T^Y(\omega)]e^T|HF > \\
& \frac{1}{2}P(XY)\bar{\mathbf{Q}}_N^X(-\omega) \cdot \bar{\mathbf{V}}_N^Y(\omega)
\end{aligned}
\qquad (3.25)
$$

Here, $H_N = H_N(0) + \bar{\mathbf{Q}}_N^{(0)} \cdot \mathbf{V}$ is the unperturbed Hamiltonian of the molecular solute; $\bar{\mathbf{Q}}_N^X(-\omega)$ and $\mathbf{V}_N^Y(\omega)$ are given by

$$
\begin{aligned}
\bar{\mathbf{Q}}_N^X(-\omega) =\ & < HF|(1 + \Lambda)e^{-T}[\mathbf{Q}_N^{|\omega|}, T^X(-\omega)]e^T|HF > \\
& + < HF|\Lambda^X(-\omega)e^{-T}\mathbf{Q}_N^{|\omega|}e^T|HF >
\end{aligned}
\qquad (3.26a)
$$

$$
\begin{aligned}
\bar{\mathbf{V}}_N^Y(\omega) =\ & < HF|(1 + \Lambda)e^{-T}[\mathbf{V}_N, T^Y(\omega)]e^T|HF > \\
& + < HF|\Lambda^Y(\omega)e^{-T}\mathbf{V}_N e^T|HF >
\end{aligned}
\qquad (3.26b)
$$

---

[8] The variational criteria at the zero and first order are satisfied when the Coupled-cluster wave-functions are solution of the PCM coupled-cluster Eqs. (1.24) and (1.28).

In Eq. (3.26a) the polarization charges operator $\mathbf{Q}_N^{|\omega|}$ is defined as

$$\mathbf{Q}_N^{|\omega|} = \mathbf{T}(\omega) \cdot \mathbf{V}_N \tag{3.27}$$

where $\mathbf{T}(\omega)$ is the frequency dependent PCM response matrix evaluated with the dielectric permittivity $\varepsilon(\omega)$ of the solvent at the frequency ω (see Appendix A.3).

- The stationary condition $\delta G^{XY}(-\omega, \omega) = 0$ with respect to the first order $\Lambda^X(\pm\omega)$ amplitudes give the equations for the first order $T^X(\pm\omega)$ amplitudes:

$$\begin{aligned}
0 = &< HF|\tau_p^\dagger e^{-T} \tilde{Y}(\omega) e^T |HF > \\
&+ < HF|\tau_p^\dagger [e^{-T} H_N e^T, T^Y(\omega)] - \omega T^Y(\omega)|HF >
\end{aligned} \tag{3.28}$$

where $\tilde{Y}(\omega)$ is an effective perturbing operator:

$$\tilde{Y}(\omega) = Y + \bar{\mathbf{Q}}_N^Y(\omega) \cdot \mathbf{V}_N \tag{3.29}$$

Here, the term "effective" indicates that in addition to the direct perturbation (Y), there is an indirect source of perturbation, $\bar{\mathbf{Q}}_N^Y(\omega) \cdot \mathbf{V}_N$, due to the coupling between the direct perturbation and the solute-solvent interaction.
- The stationary condition $\delta G^{XY}(-\omega, \omega) = 0$ with respect to the first order $T^X(\pm\omega)$ amplitudes give the equations for the first order $\Lambda^X(\pm\omega)$ amplitudes:

$$\begin{aligned}
0 = &< HF|(1 + \Lambda) e^{-T} [\tilde{Y}(\omega), \tau_p] e^T |HF > \\
&+ < HF|(1 + \Lambda) e^{-T} [[H_N, \tau_p], T^Y(\omega)] e^T |HF > \\
&- \omega < HF|\Lambda^Y(\omega)\tau_p|HF > + < HF|\Lambda^Y(\omega) e^{-T} [H_N, \tau_p] e^T |HF >
\end{aligned} \tag{3.30}$$

3.2.2 Coupled-Cluster Linear and Quadratic Response Functions

- The linear response functions $<< X; Y >>_\omega$ can then be obtained from Eq. (3.24) by introducing the stationary condition (3.28) into the second order quasi-free-energy (3.25)

$$\begin{aligned}
<< X; Y >>_\omega = &\tilde{G}^{XY} \\
= &\tfrac{1}{2} C^{\pm\omega} \{ P(XY) < HF|(1 + \Lambda) e^{-T} [[X, T^Y(\omega)] e^T |HF > \\
&+ < HF|(1 + \Lambda) e^{-T} [[H_N, T^X(-\omega)], T^Y(\omega)] e^T |HF > \\
&\tfrac{1}{2} P(XY) < HF|(1 + \Lambda) e^{-T} [\mathbf{Q}_N^{|\omega|}, T^X(-\omega)] e^T |HF > \\
&\cdot < HF|(1 + \Lambda) e^{-T} [\mathbf{V}_N, T^Y(\omega)] e^T |HF > \\
&\tfrac{1}{2} P(XY) < HF|\Lambda^X(-\omega)\mathbf{Q}_N^{|\omega|} e^T |HF > \cdot < HF|\Lambda^Y(\omega)\mathbf{V}_N e^T |HF > \}
\end{aligned} \tag{3.31}$$

where $C^{\pm\omega}$ is a symmetrization operator with respect to sign change of the frequencies to ensure that only the real part of the response function is retained, and $P(x, y)$ is a permutation operator between x and y.

Alternatively, introducing the stationary condition (3.30) in to the second order quasi-free-energy (3.25) we can write the linear response functions as

$$
<< X; Y >>_\omega = \tfrac{1}{2} C^{\pm\omega} \{ < HF|(1+\Lambda)e^{-T}[X, T^Y(\omega)]e^T|HF > \\
+ < HF|\Lambda^Y(\omega)e^{-T} X e^T|HF > \}
\tag{3.32}
$$

- The quadratic response functions $<< X; Y, Z >>_{\omega_Y,\omega_Y}$ can be obtained in agreement with the $(2n+1)$ rule[9] as:

$$
<< X; Y, Z >>_{\omega_Y,\omega_Y} = \tilde{G}^{XYZ}
$$
$$
= \frac{1}{2} C^{\pm\omega} \{ P^3(XYZ) < HF|(1+\Lambda)e^{-T}|[[X, T^Y(\omega_Y), T^Z(\omega_Z)]e^T|HF >
$$
$$
+ P^6(XYZ) < HF|\Lambda^X(\omega_X)e^{-T}[Y, T^Z(\omega_Z)]e^T|HF >
$$
$$
+ < HF|(1+\Lambda)e^{-T}[[[H_N, T^X(\omega_X)], T^Y(\omega_Y)], T^Z(\omega_Z)]e^T|HF >
$$
$$
+ P^3(XYZ) < HF|\Lambda^X(\omega_X)e^{-T}[[H_N, T^Y(\omega_Y)], T^Z(\omega_Z)]e^T|HF >
$$
$$
\frac{1}{2} P^3(XYZ) \left(\tilde{\mathbf{Q}}_N^{XY}(\omega_X+\omega_Y) \cdot \bar{\mathbf{V}}_N^Z(\omega_Z) + \bar{\mathbf{Q}}_N^X(\omega_X) \cdot \tilde{\mathbf{V}}_N^{YZ}(\omega_Y+\omega_Z) \right) \}
\tag{3.33}
$$

where $P^3(xyz)$ and $P^6(xyz)$ denotes, respectively the cyclic permutation operator and the full permutation operator of their arguments, while $\tilde{\mathbf{Q}}_N^{XY}(\omega_X+\omega_Y)$[10] and $\tilde{\mathbf{V}}_N^{YZ}(\omega_Y+\omega_Z)$ are defined as

$$
\tilde{\mathbf{Q}}_N^{XY}(\omega_X+\omega_Y) = < HF|(1+\Lambda)e^{-T}[[\mathbf{Q}_N^{|\omega_X+\omega_Y|}, T^X(\omega_X), T^Y(\omega_Y)]e^T|HF>
$$
$$
+ P(XY) < HF|\Lambda^X(\omega_X)e^{-T}[\mathbf{Q}_N^{|\omega_Z|}, T^Y(\omega_Y)]e^T|HF >
\tag{3.34a}
$$
$$
\tilde{\mathbf{V}}_N^{YZ}(\omega_Y+\omega_Z) = < HF|(1+\Lambda)e^{-T}[[\mathbf{V}_N, T^X(\omega_X), T^Y(\omega_Y)]e^T|HF >
$$
$$
P(XY) < HF|\Lambda^X(\omega_X)e^{-T}[\mathbf{V}_N, T^Y(\omega_Y)]e^T|HF >
\tag{3.34b}
$$

In the zero-frequency limit, the linear response functions (3.32) give, respectively, the second derivative of the free-energy functional (2.25) and (2.31). Similarly, the quadratic response function (3.33) gives, in the zero-frequency limit the third-derivative of the time-independent free energy functional (2.32).

3.2.2.1 The Response Functions in the PTE Approximation

In the PCM-CC-PTE approximation, i.e. in the presence of the fixed HF reaction field. The response functions are still given by Eqs. (3.31), (3.32) and (3.33), respectively.

---

[9] Third- and second-order derivatives of the T, Λ amplitudes are eliminated, respectively, by the zero- and first order stationarity of the quasi-free-energy functional ΔG_{CC}.

[10] The polarization charges operator $\mathbf{Q}_N^{|\omega_X+\omega_Y|}$ is evaluated from Eq. (3.27) using the dielectric permittivity $\varepsilon(|\omega_X+\omega_Y|)$ of the solvent at the frequency $|\omega_X+\omega_Y|$. Therefore, Eq. (3.34a) is able to describe the non-equilibrium solvation effects in the quadratic response function describing general second-order molecular processes [4] (see Appendix B in Ref. [19]).

However, the zero order T and Λ amplitudes to be used are those determined from the PCM-PTE Eqs. (1.30a) and (1.30b), while the corresponding first order T^Y and Λ^Y amplitudes must be obtained from modified first-order Eqs. (3.28) and (3.30), respectively. The modified first-order response equations are obtained by substituting the effective Hamiltonian H_N of Eq. (1.25) appearing in these equations with the PCM-PTE Hamiltonian $H(0)_N$.

3.3 Effective Response Functions and Macroscopic Susceptibilities

The molecular response functions defined in the previous section correspond to changes of the properties of a molecular solute when it is subjected to perturbing electric or magnetic fields whose amplitudes are defined within the cavity hosting the solute. However, these cavity fields are not measurable quantities and the molecular response functions cannot be directly connected with their experimental counterpart, the macroscopic susceptibilities. The macroscopic susceptibilities are response function describing the changes of the the properties of macroscopic portion of the medium with respect to the electromagnetic field in the medium, also called Maxwell field (see Appendix A.4). In the presence of the cavity boundary the Maxwell field is modified and the effective field acting on the molecular solute, i.e. the cavity field, include these modifications. The problem to connect the Maxwell field to the cavity field is known as the "local field" problem. The solution that has been proposed within the PCM model has lead to the definition of the so called effective response properties [30].

The effective response theory describes the variation of the properties of the molecular solute when subjected to an external perturbing field which correspond to the Maxwell electromagnetic field in the medium. The resulting molecular response properties, represent effective response properties which can be directly related to macroscopic observables, after a proper averaging over the orientational states of the molecular solutes [31–33].

The effective Hamiltonian of the molecular solute M in the presence of a Maxwell field $\mathbf{E}$ in the dielectric medium can be written as

$$H = H^0 + \bar{\mathbf{Q}}(\Psi; t) \cdot \mathbf{V} + V'(\mathbf{E}, t) \tag{3.35}$$

where H^0 and $\bar{\mathbf{Q}}(\Psi; t) \cdot \mathbf{V}$ have the same meaning as in Eq. (4.2), and $V'(\mathbf{E}, t)$ denotes the effective operator describing the interaction of the molecular solute with a Maxwell field $\mathbf{E}$, uniform far from the cavity C hosting the molecular solute.

In the case of a monochromatic field $\mathbf{E}(\omega)$ of frequency ω and uniform far from the cavity effective perturbation $V'(\mathbf{E}, t)$ can be written in terms of the electric-dipole interaction with $\mathbf{E}(\omega)$, and in terms of the interaction with an additional set of apparent surface charges $\tilde{\mathbf{Q}}^{\mathbf{E}}$ spread on the cavity surface Γ:[11]

---

[11] The contribution is due to the effect of the boundary condition of the cavity on the Maxwell field.

$$V'(\mathbf{E}, t) = -\mathbf{E} \cdot (\hat{\mu} + \mathbf{m})(e^{-i\omega t} + e^{+i\omega t}) \tag{3.36}$$

Here, $\hat{\mu} = \sum_i^N \mathbf{r}_i$ is the usual electric dipole operator, and $\mathbf{m}$ is a vectorial operator having Cartesian components:

$$m_i = \tilde{\mathbf{Q}}_i^{\mathbf{E}} \cdot \mathbf{V} \tag{3.37}$$

where, $\tilde{\mathbf{Q}}_i^{\mathbf{E}}$ collects the additional apparent charges induces by the i-th Cartesian components of a Maxwell field having unitary amplitude, and $\mathbf{V}$ is the familiar vector operator of the electrostatic potential at the positions of the charges $\tilde{\mathbf{Q}}_i^{\mathbf{E}}$.[12]

Introducing Eq. (3.36), the effective Hamiltonian of the molecular solute M can be written as

$$H = H^0 + \bar{\mathbf{Q}}(\Psi; t) \cdot \mathbf{V} - \mathbf{E} \cdot \tilde{\mu}(e^{-i\omega t} + e^{+i\omega t}) \tag{3.38}$$

where $\tilde{\mu} = \mu + \mathbf{m}$ is an effective electric-dipole operator describing the effective interaction with the Maxwell perturbing field.

The response theory described in the Sect. (3.2) can be used to describe the response of the molecular solute to the macroscopic Maxwell field in the medium. More specifically, the effective response functions are the coefficients of a Fourier-perturbation expansion of the time-dependent expectation values of a generic observable X, $< X(t) >=< \Psi(t)|X|\Psi(t) >$ of the molecular solute in order of the perturbing Maxwell field $\mathbf{E}$:

$$< \Psi(t)|X|\Psi(t) > = < X >_0 + \sum_{j,y} e^{(-i\omega_i t)} << X; \tilde{Y} >>_{\omega_j} \varepsilon_y(\omega_i)$$

$$+ \frac{1}{2} \sum_{i,j,y,z} e^{(-i(\omega_i + \omega_j)t)} << X; \tilde{Y}, \tilde{Z} >>_{\omega_y + \omega_z} \varepsilon_y(\omega_i)\varepsilon_z(\omega_j) + \cdots \tag{3.39}$$

where $<< X; \tilde{Y} >>_{\omega_Y}$ represents the effective linear-response function describing the contribution of $X(t)$ of first-order in the effective perturbation $\tilde{Y}$ with frequency ω_Y, $<< X; \tilde{Y}, \tilde{Z} >>_{\omega_Y, \omega_Z}$ denotes the quadratic-response function describing the contribution of $X(t)$ quadratic in the effective perturbations $\tilde{Y}, \tilde{Z}$ with frequencies ω_Y, ω_Z, and in a similar way are defined the higher-order response functions $<<$

---

[12] The charges $\tilde{\mathbf{Q}}_i^{\mathbf{E}}$ are obtained as solution of the electrostatic problem (i.e. the Laplace problem) describing the Maxwell field $\mathbf{E}$ in the presence of the void cavity. The corresponding integral equation with domain on the PCM cavity boundary Γ is:

$$\left(2\pi \frac{\epsilon + 1}{\epsilon - 1} + D^*\right)\sigma_i^E(\mathbf{s}) = -E_i n_i(\mathbf{s}) \; \mathbf{s} \subset \Gamma$$

where E_i and n_i are, respectively the i-th Cartesian component of the Maxwell field $\mathbf{E}$ and of a unit vector $\mathbf{n}(r)$ normal to the cavity surface at the point $\mathbf{s}$, and $\sigma_i^E(\mathbf{s})$ is the apparent surface charge density determined by the Maxwell field.

The discretization of the the apparent surface charge density $\sigma_i^E(\mathbf{s})$ leads then to the discrete set of charges $\tilde{\mathbf{Q}}_i^{\mathbf{E}}$.

Table 3.2 Examples of effective response functions properties and of the corresponding macroscopic susceptibilities

Effective response function	Effective molecular properties	Macroscopic susceptibility
$<< X; \tilde{Y} >>_{\omega=0}$ [31]	Static polarizability, $\tilde{\alpha}_{XY}(0; 0)$	Static dielectric constant, $\varepsilon(0)$
$<< X; \tilde{Y} >>_{\omega}$ [11–13, 32]	Dynamic polarizability, $\tilde{\alpha}_{XY}(-\omega; \omega)$	First order susceptibility, $\chi^{(1)}(-\omega : \omega)$
$<< X; \tilde{Y}, \tilde{Z} >>_{\omega,0}$ [13]	First hyper-polarizability, $\tilde{\beta}_{XYZ}(-\omega; \omega, 0)$	Third-order susceptibility, $\chi^{(3)}(-\omega : \omega, 0, 0)$ Kerr constant
$<< X; \tilde{Y}, \tilde{Z} >>_{\omega,\omega}$ [11, 32]	First hyper-polarizability, $\tilde{\beta}_{XYZ}(-2\omega; \omega, \omega)$	Third-order susceptibility, $\chi^{(3)}(-2\omega : \omega, \omega, 0)$ EFISH process

X, denotes a Cartesian component of the electric dipole operator $\hat{\mu}$, while $\tilde{Y}$ denotes a Cartesian component of the effective dipole operator $\tilde{\mu}$, and $\chi^{(n)}$ denotes a macroscopic susceptibilities of nth-order

$X; \tilde{Y}, \tilde{Z}, .. >>_{\omega_Y, \omega_Z,...}$. Selected examples of effective molecular response functions and of the corresponding macroscopic susceptibilities are reported in Table 3.2.

References

1. J. Tomasi, B. Mennucci, R. Cammi, Chem. Rev. **105**, 2999 (2005)
2. J. Olsen, P. Jørgensen, in *Modern electronic structure theory*, vol. 2, ed. by D. Yarkony (World Scientific, Singapore, 1995), p. 857
3. T. Helgaker, S. Coriani, P. JørgensenK, J. Olsen Kristensen, K. Ruud, Chem. Rev. **112**, 543 (2012)
4. R. Cammi, M. Cossi, B. Mennucci, J. Tomasi, J. Chem. Phys. **105**, 10556 (1996)
5. R. Cammi, B. Mennucci, J. Chem. Phys. **110**, 9877 (1999)
6. J. Tomasi, R. Cammi, B. Mennucci, Int. J. Quantum Chem. **75**, 783 (1999)
7. B. Mennucci, J. Tomasi, R. Cammi, J.R. Cheesman, M.J. Frisch, F.S. Devlin, S. Gabriel, P.J. Stephens, J. Chem. Phys. A **106**, 6102 (2002)
8. R. Cammi, B. Mennucci, J. Tomasi, J. Phys. Chem. A **104**, 5631 (2000)
9. M. Cossi, V. Barone, J. Chem. Phys. **115**, 4701 (2001)
10. R. Cammi, L. Frediani, B. Mennucci, K., Rudd. J. Chem. Phys. **119**, 5818 (2003)
11. C. Cappelli, B. Mennucci, J. Tomasi, R. Cammi, A. Rizzo, G.L.J.A., Rikken, R. Mathevet, C. Rizzo. J. Chem. Phys. **118**, 10712 (2003)
12. C. Cappelli, B. Mennucci, R. Cammi, A. Rizzo, J. Phys. Chem. B **109**, 18706 (2005)
13. L. Frediani, Z. Rinkevicius, H. Ågren. J. Chem. Phys. **122**, 244104 (2005)
14. B. Jansík, A. Rizzo, L. Frediani, K. Ruud, S. Coriani, J. Chem. Phys. **125**, 234105 (2006)
15. A. Rizzo, L. Frediani, K. Ruud, J. Chem. Phys. **127**, 164321 (2007)
16. K. Zhao, L. Ferrighi, L. Frediani, C.-K. Wang, Y. Luo, J. Chem. Phys. **126**, 204509 (2007)
17. L. Ferrighi, L. Frediani, E. Fossgaard, K. Ruud, J. Chem. Phys. **127**, 244103 (2007)

18. N. Lin, L. Ferrighi, X. Zhao, K. Ruud, A. Rizzo, Y. Luo, J. Phys. Chem. B **112**, 4703 (2008)
19. R. Cammi, Int. J. Quantum Chem. **112**, 2547 (2012)
20. R. Cammi, Adv. Quantum Chem. **64**, 1 (2012)
21. R.A. Marcus, J. Phys. Chem. 24, 966 (1956); (b) M.D. Newton, H.L. Friedman, J. Chem. Phys. 88, 4460(1988); (c) H.J. Kim, J.T. Hynes, J. Chem. Phys. 93, 5194(1990); (d) M.V. Basilevsky, G.E. Chudinov, Chem. Phys. 144, 155 (1990); (e) M.A. Aguilar, F.J. Olivares del Valle, J. Tomasi, J. Chem. Phys. 98, 7375(1993); (f) D.G. Truhlar, G.K. Schenter, B.C. Garrett, J. Chem. Phys. 98, 5756(1993); (g) R. Cammi, J. Tomasi, Int. J. Quantum Chem. (Symposium) 29, 465(1995); (h) C.P. Hsu, X. Song, R.A. Marcus, J. Phys. Chem. B 101, 2546 (1997); (i) M. Caricato, B. Mennucci, J. Tomasi, F. Ingrosso, R. Cammi, S. Corni, G. Scalmani. J. Chem. Phys. **124**, 124520 (2006)
22. R. Cammi, J. Tomasi, Int. J. Quantum Chem. **60**, 297 (1996)
23. J. Frenkel, *Wave Mechanics: Advanced General Theory* (Clarendon Press, Oxford, 1934)
24. H. Sambe, Phys. Rev. A **7**, 2203 (1973)
25. P.W. Langhoff, S.T. Epstein, M. Karplus, Rev. Mod. Phys. **105**, 602 (1972)
26. F. Aiga, K. Sasagane, R. Itoh, J. Chem. Phys. **99**, 3779 (1993)
27. K. Sasagane, F. Aiga, R. Itoh, J. Chem. Phys. **99**, 3738 (1993)
28. F. Aiga, T. Tada, R. Yoshimura, J. Chem. Phys. **111**, 2878 (1999)
29. O. Christiansen, P. Jørgensen, C. Hättig, Int. J. Quantum Chem. **68**, 1 (1998)
30. R. Wortmann, K. Elich, S. Lebus, W. Liptay, P. Borowicz, A. Grabowska, J. Phys. Chem. **92**, 9724 (1992)
31. R. Cammi, B. Mennucci, J. Tomasi, J. Phys. Chem. A **102**, 870 (1998)
32. R. Cammi, B. Mennucci, J. Tomasi, J. Phys. Chem. A **104**, 4690 (1998)
33. R. Cammi, B. Mennucci, in *Continuum Solvation Models in Chemical Physics*, ed. by B. Mennucci, R. Cammi (Wiley, Chichester, 2007), p. 238

Chapter 4
Excitation Energies and Transition Moments from the PCM Linear Response Functions

Abstract This chapter considers the properties of the molecular solute in electronic excited states determined from the linear response functions described in the previous Chap. 3. Transition energies and transition moments are determined from a generalized eigenvalue equations, and the first-order properties in electronic excited states are expressed as analytical gradients of the corresponding transition energies with respect to suitable perturbations.

4.1 Excitation Energies and Transition Moments from the PCM-CC Linear Response Functions

The linear response functions $<< X; Y >>_{\omega_X}$ described in the previous Chap. 3 provide direct information about the transition properties of the unperturbed molecular solutes. The poles (ω_K) of $<< X; Y >>_{\omega_X}$ correspond to the transition energies from the ground state [1], while the residues determine the associated transition matrix elements. However, at variance with the case of an isolated molecule, the excitation energies from linear response functions of a molecular solute do not correspond to the excitation energies obtained as differences of the energies of the excited states described explicitly by a CI-like wavefunction expansion [1–3].

The lack of correspondence between the LR methods and the CI approaches have been analyzed by Kongsted [1] and Cammi and Corni [2, 3]. Methods based on the LR functions give the excitation energies and transition moments by solving a single eigenvalue problem for all the excited states. Moreover, the LR base methods give a descriptions of the solute-solvent interaction in the excites states which does not depend on the one-particle density of the molecular solutes. On the contrary, the methods based on a CI-like expansion of excited states lead to a state-specific approach which requires the solution of a separate CI-type eigenvalue equation for

---

[1] Here we consider the linear response function for the electronic ground state of the molecular solute.

R. Cammi, *Molecular Response Functions for the Polarizable Continuum Model*, Springer Briefs in Electrical and Magnetic Properties of Atoms, Molecules, and Clusters, DOI: 10.1007/978-3-319-00987-2_4, © The Author(s) 2013

each of the different excited states of interest. Each excited state is characterized by a different Hamiltonian, due to the explicit dependence of the solute-solvent interaction operator on the wavefunction of excited state via the corresponding one-particle density.

Within the PCM framework, coupled-cluster methods for excited states based on the CI-like expansion EOM-CC [4] and SAC-CI [5], have already been presented in Refs. [6–8] and [9, 10], respectively.

4.1.1 Poles of the PCM-CC Linear Response Functions

To determines spectral form of the PCM-CC linear response function $<< X; Y >>_\omega$ of Eq. (3.31) we combine the T^X and Λ^X first-order response Eqs. (3.28–3.30) into a single matrix response equation:

$$\left[\begin{pmatrix} \mathbf{G}^{[2],\dagger}_{t,\lambda} & \mathbf{G}^{[2]}_{t,t} \\ \hline \mathbf{G}^{[2]}_{\lambda,\lambda} & \mathbf{G}^{[2]}_{t,\lambda} \end{pmatrix} - \omega \begin{pmatrix} -\mathbf{I} & \mathbf{0} \\ \hline \mathbf{0} & +\mathbf{I} \end{pmatrix} \right] \begin{pmatrix} \lambda^X(\omega_X)^\dagger \\ t^X(\omega_X) \end{pmatrix} = \begin{pmatrix} \mathbf{X}^{[1],\dagger}_\lambda \\ \mathbf{X}^{[1]}_t \end{pmatrix} \qquad (4.1)$$

Here, ω is the frequency of the perturbation, $\lambda^X(\omega_X)^\dagger$ and $t^X(\omega_X)$ are column vectors collecting, respectively, the amplitudes of the first-order Λ^X and T^X amplitudes; $\mathbf{X}^{[1]}_\lambda$ and $\mathbf{X}^{[1],\dagger}_t$ are column vectors collecting, respectively, the matrix elements $< HF|(1+\Lambda)e^{-T}[Y,\tau_p]e^T|HF >$ and $< HF|\tau_p^\dagger e^{-T}Ye^T|HF >$; finally, the diagonal blocks $\mathbf{G}^{[2]}_{\lambda,t}$ and the out-of-diagonal blocks, $\mathbf{G}^{[2]}_{t,t}$ and $\mathbf{G}^{[2]}_{\lambda,\lambda}$ are defined:

$$\mathbf{G}^{[2]}_{\lambda,t}(p,q) = < HF|\tau_p^\dagger \left[e^{-T}H_N e^T, \tau_q \right]|HF > +\mathbf{Q}^{[1]}_{\lambda_p} \cdot \mathbf{V}^{[1]}_{t_p} \qquad (4.2)$$

$$\mathbf{G}^{[2]}_{t,t}(p,q) = < HF|(1+\Lambda)e^{-T}[[H_N,\tau_p],\tau_q]e^T|HF > +\mathbf{Q}^{[1]}_{t_p} \cdot \mathbf{V}^{[1]}_{t_p} \qquad (4.3)$$

$$\mathbf{G}^{[2]}_{\lambda,\lambda}(p,q) = \mathbf{Q}^{[1]}_{\lambda_p} \cdot \mathbf{V}^{[1]}_{\lambda_q} \qquad (4.4)$$

with

$$\mathbf{Q}^{[1]}_{\lambda_p} = < HF|\tau_p^\dagger|e^{-T}\mathbf{Q}^{|\omega|}_N e^T|HF >$$

$$\mathbf{V}^{[1]}_{t_p} = < HF|(1+\Lambda)e^{-T}[\mathbf{V}_N,\tau_q]e^T|HF >$$

$$\mathbf{Q}^{[1]}_{t_p} = < HF|(1+\Lambda)e^{-T}[\mathbf{Q}^{|\omega|}_N,\tau_p]e^T|HF >$$

$$\mathbf{V}^{[1]}_{\lambda_q} = < HF|\tau_q^\dagger|e^{-T}\mathbf{V}_N e^T|HF >$$

$$\lim_{\omega \to \omega_K} (\omega - \omega_K) << X; Y >>_\omega = \frac{1}{2} \left(T_{of}^X T_{fo}^Y + (T_{of}^X T_{fo}^Y)^* \right) \qquad (4.14)$$

where $<< X; Y >>_\omega$ is the linear response function of Eq. (4.7), and e T_{of}^X, T_{fo}^X, are respectively, the "left" and "right" transition moments

$$T_{of}^X = < HF|L_f e^{-T} X e^T |HF > \qquad (4.15a)$$

$$T_{fo}^X = < HF|(1 + \Lambda)[e^{-T} X e^T, R_f]|HF > + < HF|\mathcal{M} e^{-T} X e^T |HF > \qquad (4.15b)$$

where $\mathcal{M} = \sum_K \mu_K \tau_K^\dagger$ is a de-excitation operator whose amplitudes are determined as solution of a perturbation independent equation.[2]

The "left" and "right" transition moments of Eqs. (4.15a, 4.15b) determine the solute-solvent interaction in the linear response excitation energies ω_f (4.11). This it can be shown by a perturbative analysis of Eq. (4.11). Let us consider as zero-order solutions the eigenvalues $\omega_K^{(0)}$ and eigenvector $R_K^{(0)}$, and $L^{(0)}$ corresponding to excited-states in the presence of the fixed reaction field of the coupled-cluster ground state. At the first order in the solvent perturbation, we can write the excitation energies as

$$\omega_f^{(1)} \simeq < HF^{(0)}|L_f^{(0)} \left[e^{-T} H_N e^T, R_f^{(0)} \right]|HF > + \mathbf{T}_{of}^Q \cdot \mathbf{T}_{fo}^V \qquad (4.16)$$

with

$$\mathbf{T}_{of}^Q = < HF|L_f^{(0)} e^{-T} \mathbf{Q}_N^{|\omega|} e^T |HF >$$

$$\mathbf{T}_{fo}^V = < HF|(1 + \Lambda) e^{-T} [\mathbf{V}_N, R_f^{(0)}] e^T |HF >$$

Here, $\mathbf{T}_{of}^Q$ denotes the left transition moment (4.15a) associated to the polarization charge operator $\mathbf{Q}_N^{\omega_K|}$, while $\mathbf{T}_{fo}^V$ denotes an approximated form of the right transition moment (4.15b), associated to the electrostatic potential operator, $\mathbf{V}$, of the molecular solute.

We note that:

---

[2] The amplitudes of de-excitation operator $\mathcal{M} = \sum_K \mu_K \tau_K^\dagger$ are determined by the perturbation independent equation

$$< HF|(1 + \Lambda)[[e^{-T} H_N e^T, \tau_K], R_f]|HF >$$
$$+ < HF|(1 + \Lambda) e^{-T}[Q_N^{|\omega|}, \tau_K] e^T |HF > \cdot < HF|(1 + \Lambda) e^{-T}[V_N, R_f] e^T |HF >$$
$$+ < HF|M[e^{-T} H_N e^T, \tau_K]|HF >$$
$$+ < HF|M e^{-T} Q_N^{|\omega|} e^T |HF > \cdot < HF|(1 + \Lambda) e^{-T}[V_N, \tau_K] e^T |HF > + \omega \mu_K = 0$$

- The first term of Eq. (4.16) corresponds to the excitation energies of the solute in the presence of the PCM-CC fixed solvent reaction potential
- the second term represents a solute-solvent contribution determined by the interaction of the right transition density of the molecular solute represented by $\mathbf{T}_{fo}^{V}$ with the polarization charges $\mathbf{T}_{of}^{Q}$ induced by the left transition density matrix.

4.2 Excited State Properties

A wide range of first-order properties of the electronic excited-states of molecules in solution can be computed in terms of the gradients of the excitation energies ω_K with respect to external or internal perturbations. In this section, we discuss the analytical theory for the gradient of the excitation energies (4.11) computed from the coupled-cluster linear response.

4.2.1 Analytical Gradients Theory of the PCM-LRCC Excitation Energies

The first derivative the excitation energies ω_f (4.11) can be determined by means of a Lagrangian formulation [13, 14], to avoids the evaluation of the first derivative of the coupled-cluster ground state T and Λ amplitudes.

The PCM-LRCC Lagrangian is defined as

$$
\begin{aligned}
F(T, \Lambda, L_f, R_f, \omega_f) = &< HF|L_f\left[e^{-T}H_N e^{T}, R_f\right]|HF> \\
&+ < HF|L_f e^{-T}\mathbf{Q}_N^{|\omega|}e^{T}|HF> \\
&\cdot < HF|(1+\Lambda)e^{-T}[\mathbf{V}_N, R_f]e^{T}|HF> \\
&+ < HF|\mathcal{L}_f e^{-T}H_N e^{T}|HF> \\
&+ < HF|(1+\Lambda)e^{-T}[H_N, \mathcal{T}_f]e^{T}|HF> \\
&+\omega_K(1- < HF|L_f R_f|HF>)
\end{aligned}
\tag{4.18}
$$

where the last term of Eq. (4.18) introduces the bi-orthonormalization condition (4.10), and $\mathcal{L}_f$ and $\mathcal{T}_f$ are linear combinations, perturbation independent, of de-excitation and excitation operators $\tau_q^{\dagger}$:

$$
\mathcal{L}_f = \sum_q \zeta_q^f \tau_q^{\dagger}, \qquad \mathcal{T} = \sum_q \xi_q^f \tau_q^{\dagger}
$$

The Lagrangian of Eq. (4.18) is required to be stationary with respect to all its arguments $T, \Lambda, L_f, R_f, \omega_f$. Tacking the derivative with respect to the excitation energy we recover the normalization condition (4.10); tacking the derivative of F with respect to the coefficients of L_f gives the right eigenvalue Eq. (4.8) for R_f, and vice-versa the coefficients of R_f gives the left eigenvalue Eq. (4.9) for

L_f; tacking the derivative of F with respect to the coefficients of $\mathscr{L}_f$ and $\mathscr{T}_f$ give, respectively, the Eqs. (1.24, 1.28) for the T and Λ amplitudes. Finally, the stationary conditions of the functional F with respect to the variations of the coupled-cluster T and Λ amplitudes lead to a sets of coupled equations[3] which has the two set of Lagrange multipliers $\mathscr{L}_f$ and $\mathscr{T}_f$ as solution.

The derivatives of the excitation energies ω_f (4.11), with respect any perturbation α, can then be expressed as derivatives of the stationary functional (4.18)[4]:

$$
\begin{aligned}
\frac{\partial F}{\partial \alpha} = \frac{\partial \omega_f}{\partial \alpha} &= \omega_f^\alpha \\
&= <HF|L_f \left[e^{-T} H_N^{[\alpha]} e^T, R_f \right]|HF> \\
&+ <HF|L_f e^{-T} \mathbf{Q}_N^{|\omega|,\alpha} e^T |HF> \\
&\quad \cdot <HF|(1+\Lambda)e^{-T}[\mathbf{V}_N, R_f]e^T|HF> \\
&+ <HF|L_f e^{-T} \mathbf{Q}_N^{|\omega|} e^T |HF> \\
&\quad \cdot <HF|(1+\Lambda)e^{-T}[\mathbf{V}_N^\alpha, R_f]e^T|HF> \\
&+ <HF|\mathscr{L}_f e^{-T} H_N^{[\alpha]} e^T |HF> \\
&+ <HF|(1+\Lambda)e^{-T}[H_N^{[\alpha]}, \mathscr{T}_f]e^T|HF>
\end{aligned}
\tag{4.19}
$$

where $H_N^{[\alpha]}$ denotes a total derivative of the PCM-CC Hamiltonian:

$$
H_N^{[\alpha]} = H_N(0)^\alpha + \bar{\mathbf{Q}}_N^\alpha \cdot \mathbf{V}_N + \bar{\mathbf{Q}}_N \cdot \mathbf{V}_N^\alpha
$$

with $\bar{\mathbf{Q}}_N^\alpha = <HF|(1+\Lambda)e^{-T}\mathbf{Q}_N^\alpha e^T|HF>$.

---

[3] The coefficients of the $\mathscr{L}_f$ and $\mathscr{T}_f$ operators are determined from the following set of equations:

$$
\begin{aligned}
0 &= <HF|L_f \left[[e^{-T} H_N e^T, \tau_p], R_f \right]|HF> \\
&+ <HF|(1+\Lambda)[e^{-T}\mathbf{Q}_N e^T, \tau_p]|HF> \cdot <HF|L_f[e^{-T}\mathbf{V}_N e^T, R_f]|HF> \\
&+ <HF|L_f[e^{-T}\mathbf{Q}_N^{|\omega|} e^T, \tau_p]|HF> \cdot <HF|(1+\Lambda)e^{-T}[[\mathbf{V}_N, \tau_p], R_f]e^T|HF> \\
&+ <HF|\mathscr{L}_f[e^{-T} H_N e^T, \tau_p]|HF> \\
&+ <HF|(1+\Lambda)e^{-T}[[H_N, \tau_p], \mathscr{T}_f]e^T|HF>
\end{aligned}
$$

and

$$
\begin{aligned}
0 &= <HF|L_f \left[e^{-T} <HF|\tau_q^\dagger e^{-T}\mathbf{Q}_N e^T|HF> \cdot \mathbf{V}_N e^T, R_f \right]|HF> \\
&+ <HF|L_f e^{-T}\mathbf{Q}_N^{|\omega|} e^T |HF> \cdot <HF|\tau_q^\dagger e^{-T}[\mathbf{V}_N, R_f]e^T|HF> \\
&+ <HF|\mathscr{L}_f e^{-T} > HF|\tau_q^\dagger e^{-T}\mathbf{Q}_N e^T|HF> \cdot \mathbf{V}_N e^T|HF> \\
&+ <HF|\tau_q^\dagger e^{-T}[H_N, \mathscr{T}_f]e^T|HF> \\
&+ <HF|(1+\Lambda)e^{-T}[> HF|\tau_q^\dagger e^{-T}\mathbf{Q}_N e^T|HF> \cdot \mathbf{V}_N, \mathscr{T}_f]e^T|HF>
\end{aligned}
$$

[4] Note that, as the PCM-CC Eqs. (1.24, 1.28), and the PCM-LR-CC Eqs. (4.8) and (4.9) are assumed to be satisfied, the Lagrangian (4.18) is equal to the excitation energy, $F = \omega_f$.

4.2.2 Properties of Excited States: MO Relaxation, and Non-equilibrium Solvation

The analytical gradients (4.19) may be applied both to the geometrical gradients, for the exploration of the excited states PES of the molecular solute, and to the first derivatives with respect to the amplitude of an external, or internal static perturbation, for the determination of the first order properties of the excited states.

However, the two cases differ both with respect to the relaxation of the molecular orbitals (MO) of the Hartree-Fock reference state of the solute, and with respect to the presence of non-equilibrium solvation effects.

- For the geometrical gradients it is mandatory to consider the effects of the MO relaxation [15]. In this case, the gradients Eq. (4.19) may be reformulated in terms of contractions of effective density matrices and of differentiated one and two-electron MO integrals:

$$
\omega_f^\alpha = \sum_{rs} f_{rs}^{PCM,\alpha} \gamma_{rs}^f + \frac{1}{2} \sum_{rstu} (q_{rs} \cdot v_{tu})^\alpha \gamma_{rs}^{0f} \gamma_{tu}^{f0} + \frac{1}{4} \sum_{rstu} < rs||tu >^\alpha \Gamma_{rstu}^f
$$

(4.20)

where $f_{rs}^{PCM,\alpha}$ and $< rs||tu >^\alpha$ are, respectively, the derivative of the PCM Fock matrix elements [6] and of the antisymmetrized two-electron integrals, in the MO basis; γ_{rs}^f and Γ_{rstu}^f are matrix elements of the one and two-particle density matrices, respectively [16] [5] γ_{rs}^{0f} and γ_{rs}^{f0} are matrix elements of the transition one-particle transition density matrix see footnote 5. The differentiated MO integrals involve derivative of the MO coefficients, which can be avoided by solving, or by exploiting the PCM-Z-vector method [17].

For the geometrical gradients it is also convenient to consider of derivatives of the excitation energies (4.11) within an equilibrium solvation regime, [18]. This implies that the excitation energies (4.11), and the perturbation independent Lagrange multipliers $\mathscr{L}_f$ and $\mathscr{T}_f$ see footnote 3 involved in the evaluation of the geometrical gradients, must be evaluated with the solvent polarization operator $\mathbf{Q}_N^{|\omega|}$ in the limit of the static dielectric permittivity (i.e. at $\omega = \omega_0$).

---

[5] The one particle density matrix elements γ_{rs}^f, γ_{rs}^{0f} and γ_{rs}^{f0} are defined as:

$$
\gamma_{rs}^f = < HF|L_f \left[e^{-T} \{\tau_p^\dagger \tau_q\} e^T, R_f \right] |HF >
$$
$$
+ < HF|(1+\Lambda)e^{-T}[\{\tau_p^\dagger \tau_q\}, \mathscr{T}_f]e^T|HF >
$$
$$
+ < HF|\mathscr{L}_f e^{-T} \{\tau_p^\dagger \tau_q\} e^T|HF >
$$
$$
\gamma_{rs}^{0f} = < HF|(1+)e^{-T}[\{\tau_p^\dagger \tau_q\}, R_f]e^T|HF >
$$
$$
\gamma_{rs}^{f0} = < HF|L_f e^{-T} \{\tau_p^\dagger \tau_q\} e^T|HF >
$$

where $\{\tau_p^\dagger \tau_q\}e^T|HF$ denote a normal ordered sequences of the creation/annihilation operators.

- For the gradients of the excitation energies with respect to external fields we can exploit a MO unrelaxed approach [19], and Eq. (4.20) reduces to the form of an expectation value of the observable X in the excited state f:

$$\omega_f^{\alpha,eq/neq} = <HF|L_f\left[e^{-T}Xe^{T}, R_f\right]|HF> + <HF|\mathscr{Z}_f e^{-T}Xe^{T}|HF>$$
$$+ <HF|(1+\Lambda)e^{-T}[X, \mathscr{T}_f]e^{T}|HF> \tag{4.21}$$
$$= \sum_{rs} \gamma_{rs}^{f} x_{rs}$$

where x_{rs} are the one-electron MO integrals for the perturbation X, and the upperscript eq/neq denotes the equilibrium/non equilibrium solvation regime. The solvation regime depends on the choice of solvent operator $\mathbf{Q}_N^{|\omega|}$ involved in the calculation of the multipliers $\mathscr{Z}_f$ and $\mathscr{T}_f$ (see footnote 3).

Accordingly to the phenomenological theory of the solvent polarization (see Appendix A.3), the non-equilibrium solvation is required to describe the properties of the excites states immediately after a fast vertical excitation/de-excitation process, while an equilibrium solvation may be used to describe the changes in the the excited states properties of the solvated chromophores after the solvent relaxation which follows a vertical excitation process [20, 21].

References

1. J. Kongsted, A. Osted, K.V. Mikkelsen, O. Christiansen, Mol. Phys. **100**, 1813 (2002)
2. R. Cammi, S. Corni, B. Mennucci, J. Tomasi, J. Chem. Phys. **122**, 101513 (2005)
3. S. Corni, R. Cammi, B. Mennucci, J. Tomasi, J. Chem. Phys. **123**, 134512 (2005)
4. H. Sekino, R.J. Bartlett, Int. J. Quantum Chem. (Symposium) **18**, 255 (1984)
5. H. Nakatsuji, Chem. Phys. Lett. **59**, 362 (1978)
6. R. Cammi, Int. J. Quantum Chem. **110**, 3040 (2010)
7. M. Caricato, J. Chem. Theo. Comp. **8**, 4494 (2012)
8. M. Caricato, J. Chem. Theo. Comp. **8**, 5081 (2012)
9. R. Cammi, R. Fukuda, M. Ehara, H. Nakatsuji, J. Chem. Phys. **133**, 024104 (2010)
10. R. Fukuda, M. Ehara, H. Nakatsuji, R. Cammi, J. Chem. Phys. **134**, 104109 (2011)
11. O. Christiansen, P. Jörgensen, C. Hättig, Int. J. Quantum Chem. **68**, 1 (1998)
12. O. Christiansen, K.V. Mikkelsen, J. Chem. Phys. **110**, 8348 (1999)
13. P.Z. Szalay, Int. J. Quantum Chem. **55**, 151 (1995)
14. S.R. Gwaltney, R.J. Bartlett, J. Chem. Phys. **110**, 62 (1999)
15. J. Gauss, in *Encyclopedia of Computational Chemistry*, vol. I, ed. by P.V.R. Schleyer (Wiley, New York, 1999), p. 617
16. J.F. Stanton, R.J. Bartlett, J. Chem. Phys. **98**, 7029 (1993)
17. R. Cammi, B. Mennucci, J. Tomasi, J. Phys. Chem. A **103**, 9100 (1999)
18. C. Cappelli, S. Corni, B. Mennucci, R. Cammi, J. Tomasi, J. Chem. Phys. **113**, 11270 (2000)
19. H. Koch, P. Jörgensen, J. Chem. Phys. **93**, 3333 (1990)
20. R. Cammi, B. Mennucci, in *Challenges and Advances in Computational Chemistry and Physics*, vol. V, ed. by M.K. Shukla, J. Leszczynski (Springer, New York, 2008), p. 179
21. B. Mennucci, C. Cappelli, C.A. Guido, R. Cammi, J. Tomasi, J. Phys. Chem. A **113**, 3009 (2009)

Appendix A

A.1 Molecular Electronic Virial Theorem for the Polarizable Continuum Model

The electronic virial theorem (EVT) plays a key role in the QM description of molecular systems [1]. For isolated molecules, the EVT consists in a relation involving the kinetics energy, the potential energies and the Cartesian forces on the nuclei of the molecular system. In passing to a molecular solute, the EVT theorem involves additional terms regarding, the electrostatic- solute-solvent interaction energy and the molecular electric field acting on the polarization charges at the boundary of the PCM cavity.

The virial theorem for the PCM model is a consequence of the variational nature of the free-energy functional G (1.10) for exact wavefunction, as it can be proved with the method of the scaling of the electronic wavefunction.

Let us consider a uniform scaling of the electronic coordinates of the exact wavefunction:

$$\Psi_\alpha(\mathbf{r}_i) = \alpha^{3N/2}\Psi(\alpha\mathbf{r}_i) \tag{A.1}$$

where α is a scaling factor (with $\alpha = 1$ for the exact wavefunction), N is the number of electrons of the molecular solutes, and $\alpha^{3N/2}$ is a normalization factor such that $< \Psi_\alpha|\Psi_\alpha > = 1$.

With the scaled wavefunction Ψ_α the corresponding free-energy functional G_α (1.10) can be written as

$$G_\alpha(\mathbf{R}, \mathbf{s}) = < \Psi_\alpha|H^o(\mathbf{R}) + \frac{1}{2} < \Psi_\alpha|\mathbf{T} \cdot \mathbf{V}(\mathbf{R}, \mathbf{s})|\Psi_\alpha > \cdot \mathbf{V}(\mathbf{R}, \mathbf{s})|\Psi_\alpha > \tag{A.2}$$

where

$$\begin{aligned}
H^o(\mathbf{R}) &= T + V(\mathbf{R}) \\
T &= \tfrac{1}{2}\sum_i^N \nabla_i^2 \\
V(\mathbf{R}) &= \sum_{i>j}\frac{1}{|\mathbf{r}_i-\mathbf{r}_j|} - \sum_I\frac{Z_I}{|\mathbf{r}_i-\mathbf{R}_I|} + \sum_{I>J}\frac{Z_IZ_J}{|\mathbf{R}_I-\mathbf{R}_J|}
\end{aligned} \tag{A.3}$$

R. Cammi, *Molecular Response Functions for the Polarizable Continuum Model*,
Springer Briefs in Electrical and Magnetic Properties of Atoms, Molecules, and Clusters,
DOI: 10.1007/978-3-319-00987-2, © The Author(s) 2013

$\mathbf{T}$ is the PCM-IEF matrix (see Eq. (1.5)) for the polarization charges, and the vectorial PCM operator $\mathbf{V}$ (see Eq. (1.5)) collecting the total molecular electrostatic potential at the positions $\mathbf{s}_i$ of the polarization charges on the cavity surface Γ; Z_I and $\mathbf{R}_I$ are, respectively, the charge and the Cartesian vector position of the nuclei.

For our purpose, it is convenient the introduce the electronic and nuclear components of the vector operator $\mathbf{V}$

$$\mathbf{V}(\mathbf{R}, \mathbf{s}) = \mathbf{V}_e(\mathbf{R}, \mathbf{s}) + \mathbf{V}_n(\mathbf{R}, \mathbf{s}) \tag{A.4}$$

having elements

$$\begin{aligned}
[\mathbf{V}_e(\mathbf{R}, \mathbf{s})]_K &= \sum_i \frac{1}{|\mathbf{r}_i - \mathbf{s}_K|} \\
[\mathbf{V}_n(\mathbf{R}, \mathbf{s})]_K &= \sum_I \frac{Z_I}{|\mathbf{R}_I - \mathbf{s}_K|}
\end{aligned} \tag{A.5}$$

Let us now introduce the following identities

$$\begin{aligned}
<\Psi_\alpha|T|\Psi_\alpha> &= \alpha^2 <\Psi|T|\Psi> \\
<\Psi_\alpha|V(\mathbf{R})|\Psi_\alpha> &= \alpha <\Psi|V(\alpha\mathbf{R})|\Psi> \\
<\Psi_\alpha|\mathbf{V}(\mathbf{R}, \mathbf{s})|\Psi_\alpha> &= \alpha <\Psi|\mathbf{V}(\alpha\mathbf{R}, \alpha\mathbf{s})|\Psi>
\end{aligned} \tag{A.6}$$

with

$$V(\alpha\mathbf{R}) = \sum_{i>j} \frac{1}{|\mathbf{r}_i - \mathbf{r}_j|} - \sum_I \frac{Z_I}{|\mathbf{r}_i - \alpha\mathbf{R}_I|} + \sum_{I>J} \frac{Z_I Z_J}{\alpha|\mathbf{R}_I - \mathbf{R}_J|} \tag{A.7}$$

and

$$\begin{aligned}
\mathbf{V}(\alpha\mathbf{R}, \alpha\mathbf{s}) &= \mathbf{V}_e(\alpha\mathbf{s}) + \mathbf{V}_n(\alpha\mathbf{R}, \alpha\mathbf{s}) \\
\mathbf{V}(\alpha\mathbf{s})_e &= \sum_i \frac{1}{|\mathbf{r}_i - \alpha\mathbf{s}_i|} \\
\mathbf{V}(\alpha\mathbf{R}, \alpha\mathbf{s})_n &= \sum_I \frac{Z_I}{|\alpha\mathbf{R}_I - \alpha\mathbf{s}_i|}
\end{aligned} \tag{A.8}$$

By substituting Eq. (A.6) in to the free energy functional G_α (A.2) we have

$$\begin{aligned}
G_\alpha(\mathbf{R}, \mathbf{s}) = \alpha^2 <\Psi|T|\Psi> &+\alpha <\Psi|V(\alpha)\mathbf{R})|\Psi> \\
&+\frac{\alpha^2}{2} <\Psi|\mathbf{T} \cdot \mathbf{V}(\alpha\mathbf{R}, \alpha\mathbf{s})|\Psi> \cdot <\Psi|\mathbf{V}(\alpha\mathbf{R}, \alpha\mathbf{s})|\Psi>
\end{aligned} \tag{A.9}$$

Let us now impose the variational condition of the exact wavefunction ($\alpha = 1$) on the free energy functional G_α of Eq. (A.9), with respect to the scaling factor α

$$\frac{\partial}{\partial\alpha} G_\alpha(\mathbf{R}, \mathbf{s})|_{\alpha=1} = 0 \tag{A.10}$$

The variational condition (A.10) lead to

$$2 <\Psi|T|\Psi> + <\Psi|V(\mathbf{R})|\Psi> + <\Psi|\mathbf{V}(\mathbf{R},\mathbf{s})|\Psi> \cdot <\Psi|\mathbf{V}(\mathbf{R},\mathbf{s})|\Psi>$$

$$= <\Psi|\frac{\partial}{\partial\alpha}V(\alpha\mathbf{R})|_{\alpha=1}|\Psi> + \frac{1}{2}\left\{<\Psi|\mathbf{T}\cdot\frac{\partial}{\partial\alpha}\mathbf{V}(\alpha\mathbf{R},\alpha\mathbf{s})|_{\alpha=1}|\Psi> \cdot <\Psi|\mathbf{V}(\mathbf{R},\mathbf{s})|\Psi>\right.$$

$$\left. + <\Psi|\mathbf{T}\cdot\mathbf{V}(\mathbf{R},\mathbf{s})|\Psi> \cdot <\Psi|\frac{\partial}{\partial\alpha}\mathbf{V}(\alpha\mathbf{R},\alpha\mathbf{s})|_{\alpha=1}|\Psi>\right\} \quad\text{(A.11)}$$

Eq. (A.11) constitutes the quantum mechanical virial theorem for molecular solutes described within the PCM model, which involves, on the left side, the kinetic and the total potential energies for exact state-wavefunctions. The terms on right side of Eq. (A.11) have a physical meaning which can be clarified with the aid of the Hellmann-Feynman theorem discussed in Chap. 2 (see Eq. 2.1).

Let us now consider for the molecular solute a perturbation corresponding to the uniform scaling of the Cartesian co-ordinates of the nuclei and of the Cartesian co-ordinates of the polarization point charges:

$$G(\alpha\mathbf{R},\alpha\mathbf{s}) = <\Psi|T + V(\alpha\mathbf{R} + \frac{1}{2}<\Psi|\mathbf{T}\cdot\mathbf{V}(\alpha\mathbf{R},\alpha\mathbf{s})|\Psi> \cdot\mathbf{V}(\alpha\mathbf{R},\alpha\mathbf{s})|\Psi> \tag{A.12}$$

By applying the Hellmann-Feynman theorem (2.1), the first derivative of the free-energy functional (A.12) with respect to the scaling factor α can be written as:

$$\frac{dG(\alpha\mathbf{R},\alpha\mathbf{s})}{\alpha}\bigg|_{\alpha=1} = <\Psi|\frac{\partial}{\partial\alpha}V(\alpha\mathbf{R})|_{\alpha=1}|\Psi>$$

$$+ \frac{1}{2}\left\{<\Psi|\mathbf{T}\cdot\frac{\partial}{\partial\alpha}\mathbf{V}(\alpha\mathbf{R},\alpha\mathbf{s})|_{\alpha=1}|\Psi> \cdot <\Psi|\mathbf{V}(\mathbf{R},\mathbf{s})|\Psi>\right.$$

$$\left. + <\Psi|\mathbf{T}\cdot\mathbf{V}(\mathbf{R},\mathbf{s})|\Psi> \cdot <\Psi|\frac{\partial}{\partial\alpha}\mathbf{V}(\alpha\mathbf{R},\alpha\mathbf{s})|_{\alpha=1}|\Psi>\right\} \quad\text{(A.13)}$$

Therefore, by introducing Eq. (A.13) into Eq. (A.11) we obtain:

$$2 <\Psi|T|\Psi> + <\Psi|V(\alpha\mathbf{R})|\Psi>$$
$$+ <\Psi|\mathbf{V}(\mathbf{R},\alpha\mathbf{s})|\Psi> \cdot <\Psi|\mathbf{V}(\mathbf{R},\mathbf{s})|\Psi> = \frac{dG_{(\alpha\mathbf{R},\alpha\mathbf{s})}}{\alpha}\bigg|_{\alpha=1} \tag{A.14}$$

Eq. (A.14) shows that the sum of two-times the kinetic energy with the total potential energy, including the solute-solvent interaction, is equal to the first derivative of the free-energy functional G with respect to a uniform scaling of the Cartesian co-ordinates of the nuclei and of the position of the polarization charges. Equation. (A.13) reduces to VT case for isolated molecules if all the solvation contribution are neglected.

It can be also be shown [2] that the right side of Eq. (A.14) can be expressed in terms of the Cartesian forces on the nuclei ($\mathbf{F}_I$), and the total molecular electric field ($\mathbf{E}_K(\mathbf{s}_K)$) at the boundary Γ of the PCM cavity hosting the molecular solute:

$$\frac{d}{d\alpha} G(\alpha \mathbf{R}, \alpha \mathbf{s}) = \sum_I \mathbf{R}_i \cdot \mathbf{F}_I + \sum_K \mathbf{s}_K \cdot \mathbf{E}_K q_K \qquad (A.15)$$

where $\mathbf{F}_I = dG(\mathbf{R})/d\mathbf{R}_I$ are Cartesian forces on the nuclei, and $\mathbf{E}_K(\mathbf{s}_K) = d[V(\mathbf{R}, \mathbf{s})]_K/d\mathbf{s}_K$ is the electric field at the position $\mathbf{s}_K$.

Introducing Eq. (A.15) into the virial theorem (A.14) we obtain a relation between the various components (kinetic and potential) of the molecular solute and the Cartesian forces on the nuclei, and the total molecular electric field at the boundary Γ of the PCM cavity hosting the molecular solute

$$\begin{aligned} & 2 < \Psi | T | \Psi > + < \Psi | V(\mathbf{R}) | \Psi > \\ & + < \Psi | \mathbf{V}(\mathbf{R}, \mathbf{s}) | \Psi > \cdot < \Psi | \mathbf{V}(\mathbf{R}, \mathbf{s}) | \Psi > = \\ & \sum_I \mathbf{R}_i \cdot \mathbf{F}_I + \sum_K \mathbf{s}_K \cdot \mathbf{E}_I q_K \end{aligned} \qquad (A.16)$$

A.2 Time-Dependent Schrödinger Equation for Nonlinear Hamiltonians

Let us consider a system characterized by the presence in the Hamiltonian of a nonlinear potential energy term, depending on the wavefunction Ψ of the system:

$$H = H^0 + V(\Psi) = -\frac{1}{2} \sum_{j=1}^{N} \nabla_j^2 + U(\mathbf{r}_i) + V(\Psi) \qquad (A.17)$$

H is the effective non-linear Hamiltonian, H^0 its linear part, and $V(\Psi)$ the non-linear component (atomic units are assumed).

We assume that $V(\Psi)$ is a functional of the first order density matrix $\rho = \rho(\Psi^* \Psi)$ having the following form:

$$V(\Psi) = \hat{A}[(\Psi^* \Psi)] \qquad (A.18)$$

where $\hat{A}$ is a suitable integral operator.

We also assume that H explicitly depends on time. This means that in general there will be an explicit dependence on time, t, in the potential energy operator: $U(\mathbf{r}, t) = Us^0(\mathbf{r}) + U'(\mathbf{r}, t)$, and that there will be an implicit time dependence in $V(\Psi)$ being Ψ function of time.

The equation of motion for this problem:

$$H\Psi = i \frac{\partial}{\partial t} \Psi \qquad (A.19)$$

may be obtained from the Hamilton principle by making use of well known techniques [3, 4]:

$$\delta I = \int_{t^0}^{t} dt' \int dr \mathscr{L} = 0 \qquad (A.20)$$

when a suitable Lagrangian density, $\mathscr{L}$, is defined.

The Lagrangian density for the nonlinear QM problem (A.19–A.20) has the form

$$\mathscr{L} = \frac{1}{2} \sum_j \nabla_j \Psi^* \cdot \nabla_j \Psi + \Psi^* \Psi [U + \frac{1}{2} V(\Psi)] - \frac{i}{2} [\Psi^* \dot{\Psi} - \dot{\Psi}^* \Psi] \qquad (A.21)$$

In fact, by introducing in the Euler-Lagrange equations associated to the Hamilton principle (A.20) we have:

$$\frac{\partial}{\partial \Psi} \mathscr{L} - \sum_{k=1}^{3N} \frac{\partial}{\partial x_k} \frac{\partial \mathscr{L}}{\partial (\partial \Psi / \partial x_k)} - \frac{\partial}{\partial t} \frac{\partial \mathscr{L}}{\partial \dot{\Psi}} = 0 \qquad (A.22)$$

$$\frac{\partial}{\partial \Psi^*} \mathscr{L} - \sum_{k=1}^{3N} \frac{\partial}{\partial x_k} \frac{\partial \mathscr{L}}{\partial (\partial \Psi^* / \partial x_k)} - \frac{\partial}{\partial t} \frac{\partial \mathscr{L}}{\partial \dot{\Psi}^*} = 0 \qquad (A.23)$$

with the Lagrangian density given by Eq. (A.21), one obtains the following expressions for the separate terms of Eq. (A.22):

$$\frac{\partial}{\partial \Psi} \mathscr{L} = \frac{\partial}{\partial \Psi} [U \Psi^* \Psi + \frac{1}{q+1} V(\Psi) \Psi^* \Psi + \frac{i}{2} \Psi \dot{\Psi}^*]$$

$$= U \Psi^* + V(\Psi) \Psi^* + \frac{i}{2} \dot{\Psi}^* \qquad (A.24)$$

$$\sum_{k=1}^{3N} \frac{\partial}{\partial x_k} \frac{\partial \mathscr{L}}{\partial (\partial \Psi / \partial x_k)} = \frac{1}{2} \sum_{j=1}^{N} \nabla_j^2 \Psi^* \qquad (A.25)$$

(with $j = 1 + \text{int} \frac{(k-1)}{3}$), and:

$$\frac{\partial}{\partial t} \frac{\partial \mathscr{L}}{\partial \dot{\Psi}^*} = -\frac{i}{2} \dot{\Psi}^* \qquad (A.26)$$

Substituting expressions (A.24)–(A.26) in Eq. (A.22), one has:

$$-\frac{1}{2} \sum_{j}^{N} \nabla_j^2 \Psi^* + U \Psi^* + V(\Psi) \Psi^* = -i \frac{\partial}{\partial t} \Psi^* \qquad (A.27)$$

which is just the conjugate of Eq. (A.19).

A.3 Non-equilibrium Solvation: The Fourier Components of the PCM Polarization Charges

The macroscopic polarization of a molecular dielectric medium results from several processes involving all the molecular degrees of freedom: translational, rotational, vibrational and electronic, which span a very wide range of characteristic relaxation times required to reach a degree of polarization thermodynamically in equilibrium with the polarizing electric field [5]. Translational contributions have characteristic time $10^{-6}s$, rotational contributions $10^{-9}s$, vibrational contribution $10^{-12}s$ and electronic contribution $10^{-15}s$. For a static perturbing field, or when the changes of the time-dependent electric field varies slowly on the scale of the molecular motions, there is time to reach always a polarization in equilibrium with the perturbing field. In this regime, which is called equilibrium polarization (i.e. equilibrium solvation) the relation between the polarization and the electric field is phenomenologically described by the static dielectric constant (ε_0) of the medium.

On the contrary, when the time-dependent electric field varies on a time scale faster than the relaxation time of one or more molecular degrees of freedom there is not time to reach at any moment a time-dependent polarization which is in equilibrium with the electric field. In this regime, which is called non-equilibrium polarization, the actual value of polarization will also depend values of the electric field at previous time, and the relation between the polarization of a dielectric medium and the time-dependent polarizing field is phenomenologically described in terms of the whole spectrum of the dielectric permittivity as a function of the frequency ω of the oscillating electric field.

In the equilibrium polarization regime the polarization charges $\bar{\mathbf{Q}}$ (1.7) of the PCM model can be expressed as an expectation value the apparent charge operator $\mathbf{Q}$,

$$\bar{\mathbf{Q}}(\Psi) = < \Psi|\mathbf{Q}|\Psi > \tag{A.28}$$

with

$$\mathbf{Q} = \mathbf{T}(\varepsilon_0) \cdot \mathbf{V} \tag{A.29}$$

Here:

- $\bar{\mathbf{Q}}$ is a vector column collecting the polarization charges $\{q(\mathbf{s}_k)\}$
- $\mathbf{T}$ is a matrix which represents the responsive polarization of the solvent, depending on its static dielectric permittivity ϵ_0 of the medium and on the geometry Γ of the cavity hosting the solute,
- $\mathbf{V}$ is vector collecting the electrostatic potential operator (2.9) of the solute at positions $\mathbf{s}_k$:

$$[\mathbf{V}]_K = V_M(\mathbf{s}_K) \tag{A.30}$$

In the non-equilibrium polarization regime, the time-dependent polarization charges of the PCM model may be written as

$$\bar{\mathbf{Q}}(\Psi; t) = \int_0^\infty < \Psi(t - \tau)|\mathbf{Q}(\tau)|\Psi(t - \tau) > d\tau \qquad \tau = t - t' \qquad \text{(A.31)}$$

where the polarization charges operator $\mathbf{Q}(\tau)$ is defined as

$$\mathbf{Q}(\tau) = \mathbf{T}(\tau) \cdot \mathbf{V} \qquad \text{(A.32)}$$

being $\mathbf{T}(\tau)$ the time-dependent IEF-PCM response matrix of the medium.

From (A.32), the non-equilibrium polarization charges (A.31) are given as

$$\bar{\mathbf{Q}}(\Psi; t) = \int_0^\infty \mathbf{T}(\tau) \cdot \bar{\mathbf{V}}(t - \tau) d\tau \qquad \text{(A.33)}$$

where $\bar{\mathbf{V}}(t - \tau)$ is

$$\bar{\mathbf{V}}(t - \tau) = < \Psi(t - \tau)|\mathbf{V}|\Psi(t - \tau) > \qquad \text{(A.34)}$$

Further, from Eq. (A.33) we can define a frequency dependent PCM response matrix ω

$$\mathbf{T}(\omega) = \int_0^\infty \mathbf{T}(\tau) e^{-i\omega\tau} d\tau \qquad \text{(A.35)}$$

The PCM response matrix $\mathbf{T}(\omega)$ connects the Fourier components of polarization charges $\bar{\mathbf{Q}}(\Psi; t)$ with the corresponding Fourier components of the molecular electrostatic potential $\bar{\mathbf{V}}(t)$:

$$\bar{\mathbf{Q}}(\omega) = \mathbf{T}(\omega) \cdot \bar{\mathbf{V}}(\omega) \qquad \text{(A.36)}$$

Eq. (A.36) can be easily shown by Fourier decomposition of the Eq. (A.33):

$$\begin{aligned}
\int_{-\infty}^\infty \bar{\mathbf{Q}}(\omega) e^{i\omega(t)} d\omega &= \int_0^\infty \mathbf{T}(\tau) e^{-i\omega\tau} \cdot \int_{-\infty}^\infty \bar{\mathbf{V}}(\omega) e^{i\omega(t)} d\omega \\
&= \int_{-\infty}^\infty \mathbf{T}(\omega) \cdot \bar{\mathbf{V}}(\omega) e^{i\omega(t)} d\omega
\end{aligned} \qquad \text{(A.37)}$$

The frequency dependent PCM matrix $\mathbf{T}(\omega)$ has the physical meaning of the IEF-PCM response matrix for a medium in the presence of a component of molecular electrostatic potential oscillating at the frequency ω, and it can be obtained from the corresponding equilibrium matrix $\mathbf{T}$ (A.32) by substituting the static dielectric permittivity ϵ_0 with the frequency dependent dielectric permittivity $\varepsilon(\omega)$.

A.4 Maxwell Field and Macroscopic Susceptibilities

The Maxwell electric field $\mathbf{E}$ is the electric component of the macroscopic electromagnetic field in material media described by the celebrate Maxwell equations. The Maxwell field $\mathbf{E}$ is coupled to the dielectric polarization field $\mathbf{P}$, by the flowing form of the Maxwell equations:

$$\nabla \times \nabla \times \mathbf{E}(\mathbf{r}, t) = -\mu_0 \frac{\partial^2}{\partial t}(\varepsilon_0 \mathbf{E}(\mathbf{r}, t) + \mathbf{P}(\mathbf{r}, t)) \tag{A.38}$$

where the polarization field $\mathbf{P}$ describes the induced dipole per volume in the material media induced by the action of the Maxwell field $\mathbf{E}$ on the electron and nuclei of the constituting particles of the media.

The macroscopic susceptibilities are the coefficients of a series expansion of the polarization field $\mathbf{P}$ with respect to the components of the Maxwell electric field $\mathbf{E}$. Given the Maxwell field

$$\mathbf{E}(t) = \mathbf{E}^0 + \mathbf{E}^\omega(e^{-i\omega t} + e^{i\omega t}) \tag{A.39}$$

the dielectric polarization response of the medium can be expressed as

$$\begin{aligned}\mathbf{P} = &\sum_{j,y} e^{(-i\omega_i t)} \chi^{(1)}(-\omega_j; \omega_j \varepsilon_y(\omega_i) + \\ &+ \frac{1}{2}\sum_{i,j,y,z} e^{(-i(\omega_i + \omega_j)t)} \chi^{(2)} \omega_y + \omega_z \varepsilon_y(\omega_i)\varepsilon_z(\omega_j) + \cdots\end{aligned} \tag{A.40}$$

The $\chi^{(n)}$ are the nth-order susceptibilities.[1] Many of these susceptibilities are measurable quantities in experiments of linear and non-linear optics (see Table A.1).

A.5 The PCM-EOM Wavefunctions and Energy Functional

In the PCM-EOM-CC/SACCI approximation the excited electronic states are represented by a linear (CI-like) expansion build-up on the coupled-cluster wavefunction for the ground state [6–8]. We use the PTE couple-cluster wavefunction, computed in the presence of the frozen Hartree-Fock reaction field, as it leads to a more simpler and physically transparent PCM-EOM theory. The EOM-CC theory leads to a non-Hermitian eigenvalue problem with right and left eigenvalues.

The EOM-CC right wavefunction for the K-th state is defined as

$$|\Psi_K> = \mathscr{R}_K e^T |HF> \tag{A.41}$$

where $e^T |HF>$ is the couple-cluster state obtained by solving the PCM-PTE equation (See Eq. (1.30a, b)), and $\mathscr{R}_K$ is a quasi-particle excitation operator

$$\mathscr{R}_K = \mathscr{R}_{K,1} + \mathscr{R}_{K,2} + \ldots \tag{A.42}$$

$$\mathscr{R}_{K,n} = \frac{1}{n!^2} \sum_{ijkl\ldots abc\ldots} r_{ijk\ldots}^{abc\ldots}(K) a_a^\dagger a_i a_b^\dagger a_j a_c^\dagger a_k \cdots$$

The EOM-CC left wavefunction is given by

---

[1] The susceptibilities $\chi^{(n)}$ are tensors of rank $n + 1$ with $3^{(n+1)}$ components.

$$< \tilde{\Psi}_K| =< HF|\mathscr{L}_K e^{-T} \qquad (A.43)$$

where $\mathscr{L}_K$ is a de-excitation operator

$$\mathscr{L}_K = \mathscr{L}_{K,1} + \mathscr{L}_{K,2} + \dots \qquad (A.44)$$

$$\mathscr{L}_{K,n} = \frac{1}{n!^2} \sum_{ijkl\dots abc\dots} l_{ijk\dots}^{abc\dots}(K) a_i^\dagger a_a a_j^\dagger a_b a_k^\dagger a_c \cdots$$

The set ket and bra wavefunctions $\mathscr{L}_K$ and $\mathscr{R}_K$ satisfy the property of bi-orthogonality many-body systems

$$< \tilde{\Psi}_K|\Psi_L > =< HF\mathscr{L}_K|\mathscr{R}_L HF > = \delta_{KL} \qquad (A.45)$$

The PCM-EOM free energy functional, ΔG_K, for the state of interest is defined as

$$\Delta G_K =< HF|\mathscr{L}_K e^{-T} H_N(0) e^T \mathscr{R}_K|HF > +\frac{1}{2}\bar{\mathbf{Q}}_N^K \cdot \bar{\mathbf{V}}_N^K \qquad (A.46)$$

where $H(0)_N$ is the normal ordered form of Hamiltonian of the solute in presence of the frozen Hartree-Fock reaction field, and $\bar{\mathbf{V}}_N^K$ and, $\bar{\mathbf{Q}}_N^K$ are the EOM expectation values for the state K of normal ordered form of the apparent charge operator and of the electrostatic potential operator defined in Table 1.1:

$$\bar{\mathbf{Q}}_N^K =< HF|\mathscr{L}_K e^{-T}\hat{\mathbf{V}}_N e^T \mathscr{R}_K|HF > \qquad (A.47)$$

$$\bar{\mathbf{V}}_N^K =< HF|\mathscr{L}_K e^{-T}\hat{\mathbf{Q}}_N e^T \mathscr{R}_K|HF > \qquad (A.48)$$

Imposing ΔG_K (A.46) stationary with respect to the $\mathscr{R}_K$ and $\mathscr{L}_K$, amplitudes we obtain a right-hand and a left-hand eigenvalue equations:

Table A.1 Selected examples of macroscopic susceptibilities and of the corresponding optical process

Macroscopic susceptibilities	Optical process
$\chi^{(1)}(0; 0)$	Dielectric constant at zero frequency, ε_0
$\chi^{(1)}(-\omega; \omega)$	Refraction index at the frequency ω, n^ω
$\chi^{(2)}(-\omega; \omega, 0)$	Change of the refraction index induced by an external static field (Pockels susceptibility)
$\chi^{(3)}(-\omega; \omega, 0, 0)$	Change of the refraction index induced by an external static field (Kerr susceptibility)
$\chi^{(2)}(-2\omega; \omega, \omega)$	Second harmonic generation (SHG)
$\chi^{(3)}(-2\omega; \omega, \omega, 0)$	Electric field induced Second harmonic generation (EFISHG)

$$\mathscr{H}_K \mathscr{R}_K | H F \rangle = \Delta E_K \mathscr{R}_K | H F \rangle \tag{A.49}$$

$$\langle H F | \mathscr{L}_K \Delta E_K = \langle H F | \mathscr{L}_K \mathscr{H}_K \tag{A.50}$$

where $\mathscr{H}_K$ is the similarity transformed Hamiltonian for the molecular solute in the Kth excited

$$\mathscr{H}_K = e^{-T} H_N^K e^T \tag{A.51}$$

$$H_N^K = H_N(0) + \bar{\mathbf{Q}}_N^K \cdot \mathbf{V}_N \tag{A.52}$$

The first term of Eq. (A.52) correspond to the Hamiltonian in the presence of the Hartree-Fock polarization charges (see Eq. (1.24)), while the second term, $\bar{\mathbf{Q}}_K^{EOM} \cdot \mathbf{V}_N$, represents the interaction of the solute with the polarization charges produced by the solute in the excited state K-th.

The PCM-EOM energy functional of Eq. (A.46) corresponds to the excitation energyState specific excitation energies from the PCM-CC ground state[2] to the K-th excited states written as sum of two contributions. For a comparison with the the corresponding expression (4.55) of the excitation energy from the PCM-CC-LR theory we note that:

- The first term of (A.46) corresponds to the vertical excitation of the solute in the presence of the fixed ground state solvent reaction potential, as in the case of the PCM-CC-LR theory
- The second term of (A.46) represents the solute-solvent contribution due to the interaction with the polarization charges induced by the one-particle density matrix of the K-th excited states. Here the difference with the corresponding term in the LR approach (4.8, 4.9) is clearly evident.
- The evaluation of the excitation energy (A.46) requires the solution of a state specific eigenvalue problem (A.49) or (A.50). Instead, the PCM-LRCC approach requires the solution of a unique eigenvalue problem (4.8) or (4.9) for all the excited states.

References

1. G. Marc, W.G. McMillan, Adv. Chem. Phys. **58**, 211 (1983)
2. R. Cammi, manuscript in preparation
3. P.M. Morse, M. Fershbach, *Methods of Theoretical Physics* (McGraw Hill, New York, 1953)
4. B.L. Moiseiwitcsh, *Variational Principles* (Interscience, London, 1966)
5. C.J.F. Böttcher, P. Bordewijk, Theory of electric polarization, (Elsevier. Amsterdam **II**, 1–4 (1978))
6. R. Cammi, Int. J. Quantum Chem. **110**, 3040 (2010)
7. R. Cammi, R. Fukuda, M. Ehara, H. Nakatsuji, J. Chem. Phys. **133**, 024104 (2010)
8. R. Fukuda, M. Ehara, H. Nakatsuji, R. Cammi, J. Chem. Phys. **134**, 104109 (2011)
9. M. Caricato, J. Chem. Theor. Comp. **8**. 4494 (2012)
10. M. Caricato, J. Chem. Theor. Comp. **8**. 5081 (2012)

---

[2] (in the PTE approximation).

Index

A

Ab-initio methods, 1
Analytical gradients (PCM-LRCC), 42
Apparent charges operator, 4, 9
Apparent surface charges (ASC), 3
Apparent surface charges method (ASC), viii
Atomic orbitals (AO), 6

B

Born–Oppenheimer approximation, 1

C

Cavity, 2–4
Cavity boundary, 3
Continuum solvation models (QM/CSM),, viii
COSMO method, viii
Coupled cluster Bruckner double (BD), 10
Coupled cluster linear response equations, 30
Coupled cluster linear response functions, 24
Coupled cluster method (CC), 6, 8, 20, 30, 31
Coupled cluster quadratic response functions, 31
Coupled cluster response theory, 27
Coupled cluster single-double (CCSD), 10
C-PCM method, viii

D

Density functional theory (DFT), 24
Dielectric permittivity, 1, 4, 52
Discrete solvation models (QM/MM), vii, ix

E

Effective electric dipole moment, 14
Effective electronic Hamiltonian, 1
Effective molecular response functions, 35
Electric dipole moment, 14

Electric dipole polarizability, 28
Electric polarizability, 14
Electron correlation, 8, 10
Electronic virial theorem, 15
Equation of Motion coupled cluster method (EOM-CC), 38, 54
Eulero-Lagrange equation, 51

F

First hyper-polarizability, 35
Forces on the nuclei, 14
Free-energy functional, 5, 6, 10, 11, 47, 49
Frequency dependent dielectric permittivity, 25, 53

G

Generalized Born method (QM/GB), viii

H

Hamilton principle, 50, 51
Harmonic vibrational frequencies, 14
Hartree–Fock method (HF), 6, 8
Hellmann–Feynman theorem, 13, 20, 49

I

IEF-PCM equation, 3
Infrared intensities, 14
Integral equation formalism (IEF-PCM), viii, 3

L

Lagrangian density, 51
Linear response coupled cluster excitation energies (PCM-LRCC), 42
Local-field, 33